David Lampe
    Dec. '8

THE CONQUEST
OF THE
NORTH ATLANTIC

THE CONQUEST
OF THE
NORTH ATLANTIC

G. J. MARCUS

New York
OXFORD UNIVERSITY PRESS
1981

© 1980 by G. J. Marcus

First published in Great Britain in 1980 by Boydell & Brewer, Ltd.
First published in the United States in 1981 by Oxford University Press, Inc.

ISBN 0-19-520252-X

LIBRARY OF CONGRESS
CATALOG CARD NO.: 80-84932

Marcus, G. J.
  The Conquest of the North Atlantic.

New York: Oxford University Press
224 p.
8102   801120

Printed in the United States of America

Contents

I. THE IRISH PIONEERS

II. THE OCEAN TRAFFIC OF THE NORTH

III. THE OCEAN VOYAGES OF THE ENGLISH AND HANSE

List of Illustrations

List of Maps

List of Abbreviations

B.L.	British Library
Bps.	*Byskupa sögur*
C.C.R.	*Calendar of Close Rolls*
C.P.R.	*Calendar of Patent Rolls*
C.P.S.	*Calendar of State Papers*
D.I.	*Diplomatarium Islandicum*
D.N.	*Diplomatarium Norwegicum*
E.	Exchequer K.R., Customs Accounts
E.H.R.	*English Historical Review*
Fær.s.	*Færeyinga saga*
Fas.	*Fornaldursögur Norðurlanda*
Fb.	*Flateyjarbók*
Fms.	*Formanna sögur*
G.H.M.	*Grønlands historiske Mindesmærke*
Gs.	*Grænlendinga saga*
Hb.	*Hauksbók*
H.C.A.	High Court of Admiralty
H.R.	*Hanserecesse*
H.U.	*Hansisches Urkundenbuch*
Íb.	*Íslendingabók*
I.E.R.	*Irish Ecclesiastical Record*
Í.F.	*Íslenzk Fornrit*
Isl.Ann.	*Islandske Annaler*
Ísl.sög.	*Íslendinga sögur*
Ksk.	*Konungs Skuggsjá*
Lb.	*Landnámabók*, ed. F. Jónsson
L.P.F.D.	*Letters and Papers Foreign and Domestic of the Reign of Henry VIII*
M.M.	*Mariner's Mirror*
M.o.G.	*Meddelelser om Grønland*
N.G.L.	*Norges gamle Love indtil 1387*
Orks.	*Orkneyinga saga*
P.R.O.	Public Record Office
R.C.	*Revue celtique*
S.G.S.	*Scottish Gaelic Studies*
Sturls.	*Sturlunga saga*

TO BJÖRN THORSTEINSSON

Acknowledgments

MY GRATEFUL THANKS are due to the following (a number of whom, I am sorry to say, are no longer living) for advice and assistance on particular points: Dr K. C. Andrews, Mr R. W. Barber, Dr Alan Binns, Dr Nora Chadwick, Mr Michael Devenish, Professor Bruce Dickens, Dr David Dumville, Professor P. G. Foote, Dr G. M. Gathorne-Hardy, Skipper Patrick Gill, Professor Jón Jóhannesson, Jacob Lund, Dr Therkel Matthiassen, Professor Sigurdur Nordal, Dr Olaf Olsen, Dr Richard Perkins, Dr Aage Roussel, Professor Hakon Shetelig, Professor Johan Schreiner, Dr Haraldur Sigurdsson, Captain Carl V. Sölver, Professor Einar Ól. Sveinsson, Skipper, J. H. Tonkin, Professor G. E. O. Turville-Petre, Dr Christian Vebæk, Mrs Elizabeth Wiggans, Commander Alan Villiers, R.N.R., Dr J. A. Williamson.

My thanks are likewise due to the Staffs of the Bodleian Library, the British Library, the National Maritime Museum, the Public Records Office, the University of London Library, and the University of Bergen Library.

I should also like to thank the Editors of the *Economic History Review*, the *English Historical Review*, the *Saga-Book of the Viking Society*, and the *Mariners Mirror* for permitting me to make use of material which appeared in these journals.

Preface

IT WOULD BE difficult to say exactly how this book began. Actually, like Topsy, it may be said to have 'growed'. So far as I can remember my interest in the progress of navigation in the North Atlantic stemmed originally from certain experiences long ago in Cornwall and the west of Ireland. At any rate, when I came to take up maritime history in after years, I was inclined to approach the problems of seamanship and navigation in the medieval era from the point of view of a coasting or fishing skipper, rather than that of the academic.

It must be well over forty years since I first made the acquaintance of the Irish curach, or sea-going canoe. Out in the Aran Islands, on a fine winter day, I crossed Foul Sound in a curach. It was an experience that I shall never forget. The sea that day was by no means so high as in the classic description of such a passage in J. M. Synge's *The Aran Islands*; the wind was light, and there were no heavy breaking seas; but a long western swell was rolling in through Foul Sound, and the curach in a seaway is like no other vessel afloat. In the ensuing years I came to know the curach and its capabilities pretty well.

Later on, while I was lecturing on maritime history during the War, I began to reflect on what may be surely regarded as one of the most intriguing problems in the annals of the sea. It is one which for long has gone unresolved. Before the astrolabe, quadrant, or mariner's compass, even, came into use in the North, the Norsemen of the Viking age were voyaging across immense stretches of open sea. How did they do it?

Books by the hundred and articles by the thousand, I suppose, have appeared treating of other aspects of the Viking age. The sagas in which many of these matters are recorded have been the subject of continuous and intensive study. Individuals and learned societies alike have been at pains to extol the literary achievement of the era. That the same Norsemen were, perhaps, the greatest seafaring folk that the world has ever known would appear almost to have escaped their notice. If a fraction of all the research and argument expended on philology and runology has been spared to remedy this omission a number of very important questions would have not have gone for so long unanswered. As it is, only in comparatively recent times have belated efforts been made to bridge the gap.

The riddle of Norse navigation interested me greatly; but I then regarded it as insoluble. I continued in that belief until, a few years later, I received a long letter from a Danish master mariner which set out, in brief but illuminating outline, a possible solution to the problem. I discovered, as I indeed expected, that the explanation prompted by his seaman's instinct and experience was entirely borne out by the evidence. That was the beginning of a close friendship and continual exchange of ideas with the late Captain Carl V. Sölver, which lasted until his death in 1966.

In most general histories the maritime aspect of affairs is by far the most neglected side of the matter. The truth is that the academic authority shies away from the difficulties and dangers of a whole range of problems that demand highly specialized knowledge to which as a rule he does not attain. The protracted and for the most part pointless controversy that arose some years ago concerning the authenticity of the 'Vinland Map' was a notable example of academic ignorance of the 'sea affair'. But for this ignorance, the pretensions of the 'Vinland Map' would surely have been rejected out of hand. As it was, a great deal of time and energy was expended to little or no effect.

The fundamental weakness of the academic approach in this field is that the savant, however erudite, is for the most part totally lacking in practical experience and understanding. For example, it is the scholar, and not the seaman, who is apt to overlook the important fact that the Pole-star is invisible in the high latitudes in summer. It was a scholar who was responsible for the staggering statement that the mariners of the medieval era would first observe the meridian altitude of the sun with an astrolabe, and then proceed to work out their longitude with an hour-glass. The scholar, relying overmuch on the journal attributed to Christopher Columbus, is inclined unduly to underrate Columbus as a navigator. The sailor, judging principally by results, will arrive at a very different conclusion. Many cases of this kind might be cited. In short, as to what is and what is not possible at sea, the mariner must necessarily be the ultimate authority.

The academic approach to these matters is admirably illustrated in certain characteristic pronouncements of the late Professor E. G. R. Taylor, who, regardless of the fact that at no time in her career was she capable of getting a rowing-boat across the Round Pond, would not hesitate to pass judgment on the professional competence of some of the greatest mariners in history. Revelling as she did in mathematics, she was apparently unable to envisage a situation wherein these problems had to be solved, and in fact were solved, without the aid of any such knowledge. On the theoretical side of the subject, her work is sound and accurate. It is on the practical side that there is a notable falling-off. There is, after all, much in maritime history which can be learned only from seamen and the sea, and not from scholars and 'book-larning'. Thus she shows very little understanding of all that is implied in the phrase 'adventitious aids to navigation'. Yet these same adventitious aids constitute an important factor in early navigation.

As a further instance of the remoteness of the academic tribe from the realities of 'the sea affair', mention might be made of another of the lady navigators of London University, who calculated that, in the fifteenth century, certain craft of Bristol, bound for Iceland, arrived at their destination at a rate of knots which seems to suggest that they must have been fitted with diesel engines; in another passage, the same authority refers to these venturers as having 'beaten a path across the trackless ocean'—surely some confusion of phraseology here. Again, at a symposium on Norse navigation staged some years ago in a certain learned journal, it transpired that the contributors did not include a single specialist in this particular field—or, for that matter, any Scandinavian authority either; or even a professional sailor. The net result of the symposium was such as might have been expected.

In one of the most interesting contributions to the subject, S. E. Morison's *The*

European Discovery of America, the author may be said to have gone to the other extreme. No armchair scholar here! It is not the Harvard Professor, indeed, but 'Admiral Sam', who immediately takes over the command with zest and aplomb. His range of interests is amazing. In a work which is primarily concerned with navigation and discovery, not infrequently he flies off at a tangent into all kinds of fascinating, if unrelated, byways—on one occasion delivering himself of something reminiscent of a State of the Union address. When he finds mere prose inadequate for his purpose he bursts into original and highly imaginative verse. The sheer exuberance of his prose carries one along at the speed of a racing yacht. In all his long life Morison never wrote a more lively, more interesting, more thoroughly entertaining work.

On the other hand, *The European Discovery of America* has the defects of its qualities. Like several other American works that could be named, it was conceived, perhaps, on too ambitious a scale. It is certainly of uneven value and authority. More than once historical truth is sacrificed to dramatic effect. There are some notable anachronisms. Morison's later chapters are markedly more effective than his earlier ones. It would appear that the further back he goes, the less assured in his grasp of the subject: for example, in his casual and sketchy survey of the early ventures into the North Atlantic of the Irish and Norsemen. These are matters which call for an intimate, specialized knowledge of the subject which the author (and, it may be added, most of his reviewers) plainly does not possess. His account of Irish maritime enterprise in the Age of the Saints is limited to a long, rambling discursive commentary on the largely legendary voyages of St. Brendan, to the neglect of such authentic and illuminating sources as the *Vita S. Columbae*. In his description of the ocean-going craft of the Norsemen he relies, almost exclusively, on the recent discoveries at Roskilde. In his account of the Norse methods of navigation, brief as it is, there are serious inaccuracies. Last but not least, in the course of his investigations into the great achievement of the Norsemen as mariners and explorers, the objection may be fairly raised that there were rather too many aeroplane flights involved and not nearly enough sober, slogging research.

What is still needed, in fact, is a very much fuller understanding of the navigation of the Western Ocean in these early centuries. It is necessary to establish clearly what is proven fact and what is no more than surmise or myth: what were the successive stages of maritime expansion to the westward, and when and how they were made; what peoples were principally concerned in these ventures into the ocean, and what were their respective contributions to the final achievement.

At the dawn of history, the great ocean marked the westernmost limits of the known world. Of what might lie beyond those apparently endless empty wastes of water, nothing was known. It is interesting to notice that certain early venturers were seemingly moved to wonder and awe at its sheer immensity. Apart from the dangers of the sea, there were all the superstitious fears engendered by myth and legend to be taken into account.

In the earliest times it was fishermen who appear to have been the first deep-watermen to brave the perils of the Western Ocean. There was, after all, much to daunt the would-be explorer. It was one of the stormiest and most hazardous regions on earth. Here was no zone of steady trade-winds or seasonal monsoons. On the contrary, the weather was constantly changing, and sometimes unpredictable. Haze, thickening to fog, was common, especially in early summer.

In gales, the low, sullen murmur of the groundswell fretting on the shore along so many hundreds of miles of coastline would rise to a menacing, thunderous roar. Once out of sight of the land, there was little to aid the mariner, apart from the celestial bodies (and that only in clear weather). Lastly—and this was a crucial factor—the prevailing winds were contrary for westward ventures. Throughout much of the year westerly predominated over easterly weather.

It is not surprising, therefore, that prehistoric man never reached the more distant Atlantic islands, as he did so many islands in other parts of the world. The earliest recorded ventures into the Western Ocean were in fact made from Ireland, and that was not until the early medieval era. It is there and then that the story begins.

Hartland
N. Devon
1980

I
THE IRISH PIONEERS

1

Curachmen of the West

FOR NEARLY AN hour the men stood waiting to launch the curach. It was blowing fresh from the south-west, with hard squalls of wind and rain; and the breakers came charging in, rank after rank, in unending succession. From time to time the men would retreat before the biggest seas, afterwards shoving their curach well out into the water again. It looked as if the long-awaited *deibhil*, or lull, would never come.

Seated in the stern of the curach, in oilskins and sou'wester, was the young priest from Inishere, the southernmost of the Aran Islands, which lay from two to three miles off across Sunda Salach (Foul Sound), at present almost invisible in the rain and spume. He had said mass early that morning after crossing over from Inishere the night before, and was now returning home.

The boat slip on Inishmaan, in the midst of the Aran group, was exposed to the full force of the heavy western swell. In anything of a breeze, the sea made fast with the weather tide. Sunda Salach was a forest of tossing wave-crests. The horizon beyond was leaden grey, and jagged like a saw. Along the black cliffs to the southward white pillars of foam were continually rising and falling. The weather showed no sign of moderating.

The usual crowd had collected down by the shore; the men in their homespuns and rawhide pampooties, the women in their red woollen gowns and petticoats, with shawls pulled over their heads. On the edge of the crowd a black dog ran to and fro, dodging the sprays and barking incessantly. A low murmur of Gaelic mingled with the hiss of the rain on the battered slip, the wailing of seabirds wheeling and circling high above the grey limestone crags, and the low, sullen thunder of the groundsea.

All this time the look-out, standing higher up the slope, was intently watching a rocky point to the southward to gauge the strength of the incoming seas. Again and again the grip of the three curachmen tightened on the gunnel of their craft as they made ready to rush her down and out into the surf; again and again the look-out, watching every sea, had waved them back.

Suddenly three great seas swept in to the shore. The men hurriedly retreated with the curach as before. Then, as a flood of yeasty grey water came swirling over the slip, the look-out snatched off his hat and waved it frantically in the air. 'Anois! Anois!' ('Now! Now!'), he shouted. The excitement of the spectators rose to fever pitch; men rushed forward to the help of the crew; together they ran the curach down the slip, plunged her into the surf—the curachmen scrambling in over the gunnel and grabbing their oars—and then, at a frenzied yell from the bowman, the helpers let go. As the next breaker bore down on them the crew almost ceased pulling: the curach's bow shot up until she nearly stood on end: a yell of apprehension arose from the crowd on the slip as a blinding cloud of spray burst over the prow: then the curach plunged safely down into the long hollow beyond, reappearing several yards further out. For a few more minutes the people stood watching the curach's rhythmic swoop up, over, and down the huge, toppling rollers. But the danger was now past and presently they turned towards their homes.

It was a scene which Inishmaan must have witnessed, time and again, for generations. Both the launching and the beaching of a curach in such weather would put the seamanship of her crew to the severest possible test. The danger of the return—if the bad weather should continue—was at least as great as that of the departure. Before the curach could be safely beached, she was in imminent peril of being swamped, or even capsized, by a heavy sea. Again the long-drawn-out suspense: the men watching and muttering among themselves, the women alternately weeping and praying, while the curach hung, for minute after anxious minute, her prow constantly turned to the sea, on the edge of the surf-line, waiting for a smooth.[1]

This unique craft, the three- or four-oared curach, perhaps the handiest, lightest, most buoyant afloat, which is still in general use in certain parts of the Atlantic coast of Ireland and nowhere else in Europe, is the descendant of one of the most ancient types of vessel in the western world. It must have been in use for centuries before the first mention of it occurs in classical literature. Both before and after the late Imperial age it was a significant factor, not only in the maritime, but also in the political and ecclesiastical, developments in these islands. It is closely knit in with the early history of the Celtic peoples. It is the key to the fuller comprehension of a whole era of seamanship and navigation. In the hide-covered craft were launched the earliest recorded ventures far out into the Western Ocean. The Irish curach has a long, a very long, history behind it.

2

Antiquity of the skin-covered craft

NOTWITHSTANDING THE paramount importance of sea-communications during the second and first millenniums B.C., very little is actually known about the shipping which plied on the western sea-routes. It is more than likely that Mediterranean craft occasionally passed the Pillars of Hercules and steered up the Atlantic coasts of western Europe. They would be relatively large, well-built vessels fitted with sails: for in the Mediterranean the sail was of great antiquity. The native shipping was apparently of two types. In calm waters such as the rivers and lakes, the Scandinavian fjords, and the sheltered reaches behind the Norwegian *skergarðr*, dug-out canoes would appear to have been generally employed. But they could hardly have been used for the passage across the northern part of the North Sea, between Cornwall, Brittany, and Ireland, or between the Nordland and the outlying isles of the Lofotens. For the open sea there is evidence, both direct and indirect, of the extensive distribution of skin-covered craft—of a type still used in the circumpolar zone—in northern and western Europe.[1]

In his *Fangstfolk* Gutorm Gjessing points to the sea-going curach of western Ireland as, in all probability, the latest descendant of the prehistoric skin craft.[2] In the course of the Bronze Age, according to another eminent authority, the late Professor A. W. Brøgger, the skin craft was 'gradually replaced by boats built of wood, cut in whole planks, with the structural details of the skin-boat as a basis'.[3] It is argued, indeed, that both the earlier skin craft and the plank-built boats of later date are depicted in the Rogaland engravings. According to Lapp tradition, skin-boats were formerly in use among their ancestors; and in comparatively recent times their plank-boats were stitched. The discovery of a number of sewn planked boats, in both Britain and Scandinavia, suggests, indeed, a widespread distribution of skin-covered craft in prehistoric times.[4]

There is little doubt that the oldest kind of sea-going craft in the high North was the skin-boat. As a result of the long and searching investigations of Gutorm Gjessing, Per and Eva Fett, J. G. D. Clark, and others, this has been traced as far back at least as the later Stone Age. Along the barren, treeless littoral of the Nordland no trace exists of either dug-out canoe or plank-boat. These, as has already been said, were craft suited for quiet waters like the inland lakes and the calm, sheltered reaches behind the *skergarðr*. Nevertheless, even in the outermost islands of the Lofoten group, the middens show that such insular settlements were in existence during the Stone Age. The middens further reveal that deep-sea fishing was carried on by the inhabitants. Finally, significant evidence is furnished by a whole series of rock-engravings situated in various parts of the Norwegian coast. Such a hull as that which so frequently occurs in these engravings would be inconceivable in a plank-built boat. The lines of the vessels depicted therein are in fact strongly suggestive of those of the skin-covered frame boats still in use among the Eskimos at the present day from Siberia eastward to Greenland. In short, the sum of the evidence serves to show that it was in substantial skin-covered vessels of the *umiak* type[5] that the inhabitants of the outlying islands during the Stone Age and afterwards journeyed to and from the Norwegian mainland and set out on their whaling and deep-sea fishing expeditions.[6]

The fragility of the skin-boat has been greatly exaggerated. The time-honoured belief that it was a frail and cranky cockle-shell of a craft probably springs from the familiar lines in the *Aeneid* concerning the coracle used by Charon—'Gemuit sub pondere cymba Sutilis, et multam accepit rimosa paludem'.[7] This impression has been strengthened, perhaps, by one's youthful memories of the famous goat-skin coracle constructed by Benn Gunn in *Treasure Island*. This is to some extent an illusion. The craft which nearly 2,000 years before Christ crossed the stormy waters of Røst could surely have been neither frail nor unseaworthy. On a number of occasions the Eskimo *kayak* has been recorded off the north of Scotland (it is possible that the Eskimo sometimes arrived in European waters even in classical times).[8]

'It was the skin-boat which made possible the settlement of all the western isles', Brøgger declares,

the further outposts of Lofoten, Vaerøy, Røst . . . In other words, there were boats which could sail or row the dangerous and famous Maelstrøm and the sea of Røst. Boats that were to be practical out there had first and foremost to be light, so that they could ride the crest of the waves and thus avoid shipping too much water. But they had also to be built strong and pliant, so that they could take the heavy stresses of the sea. Out there, a log boat would very soon break up, even if it were stabilized with outriggers. It would not be easy enough to handle, would cut through the sea and fill. Whereas the skin boat, rightly constructed, has all the advantages.[9]

The civilization of the Bronze Age was largely based upon widespread sea-communications. The evidence of archaeology is clear. The traffic embraced an immense region extending from the eastern Mediterranean to as far off as the Baltic and taking in several formidable open-sea passages. 'Only skin-boats', Bowen declares, 'would have been capable of attempting such a task'.[10] What is truly remarkable is the high level of seamanship and navigation which is implied in these ventures. To cross from Ireland and Cornwall to Brittany, and from northern Scotland to the Baltic, mariners must strike out boldly across the open sea, far out of sight of the land. The crews of the Bronze Age and earlier must have been daring and resourceful seamen. Their method of navigation is, of course, largely a matter of conjecture. One can feel fairly sure that they relied upon the sun and the stars to give them their bearings; and also that they knew something about sounding, and adventitious aids to navigation (more of this later). But of positive evidence there is not a trace. At the same time it is always to be remembered that a good deal of uncertainty surrounds the navigation of a much later epoch.

3

The curach era

IN ABOUT THE year 500 B.C. appears the first literary evidence of the hide-covered craft in the west. It occurs in a little-known passage said to have been based on a Massiliote *periplus* and also on the commentary, which is no longer extant, of the Carthaginian explorer and contemporary of Hanno, Himilco. A Greek version of this passage formed one of the sources of Festus Avienus' *Ora maritima*, which told, as a thing marvellous to relate, how the inhabitants of the southern part of what is now Brittany, who were skilled in the mining of tin and lead and were enterprising traders and seafarers, were accustomed not to fit their craft with wooden keels, 'as the ordinary usage is', but with hides joined together, and how they often 'crossed the sea in a skin'. There is a suggestion, indeed, of wonder and bewilderment in his comments on these strange and novel craft. 'Sed rei ad miraculum', exclaims Avienus,

> navigia junctis semper aptant pellibus
> corioque vastum saepe percurrunt salum.[1]

From this ancient fragment it would appear that about the middle of the first millennium B.C. reasonably strong and seaworthy hide-covered craft were plying between north-western Gaul and the 'Holy Isle', or Ireland. So, if the passage really is derived from Himilco, as has been claimed, it gives a recorded history of well over twenty-four centuries for a type of craft, constructed on the same principles, which, as we have seen, is still in use at the present time on the west coast of Ireland.

On the eve of the Roman conquest of Britain the hide-covered vessel was apparently in common use in that island. In the *De bello civile* Caesar relates that, while campaigning in Gaul, in order to transport part of his forces across a river he had a number of craft constructed of a type with which he had become acquainted in Britain some years earlier. He observes that the keel and ribs of the vessel were fashioned of light timber and the rest of the hull was wrought with wickerwork covered over with hides. It would seem from this account that the hide-covered vessels were of some size, since they had to be carried down to the river in wagons; after which the soldiers were able to cross in safety.[2]

Pliny in his *Historia naturalis* twice mentions the hide-covered craft. On the authority of Timaeus he relates that the Britons would put to sea in light craft built of wickerwork covered with hides sewn together; and he adds that they would sail these vessels to a certain island called Ictis, where 'white lead', or tin, was to be obtained.[3] In about A.D. 60 the poet Lucan describes how osiers were woven into a wickerwork hull, over which a hide was stretched. 'So sails the Venetian on the sluggish Po, and the Briton on the broad-bosomed ocean.'[4] Later, in the third century, Solinus records that hide-covered vessels of wicker (whether British or Irish is not explained) frequently crossed the Irish Sea.[5]

It was in a curach[6] that St. Patrick, who came from some region on the west coast of Britain, returned to Ireland in the fifth century (the date is very uncertain).[7] It was in a curach, too, that he must originally have been carried off by the raiders: for the *Scotti*, or Irish, came over in curachs. Somewhat later than this Sidonius Apollinaris told how the barbarian rover was accustomed to cross the sea in a curach—who, he declared, 'deems it but sport to furrow the British waters with hides, cleaving the sea in a stitched boat'.[8]

Towards the end of the Imperial era whole fleets of curachs were used in the recurrent Irish raids against Roman Britain. The worst of the raiding seems to have occurred in the years, 360, 367–8, and shortly after the turn of the century.[9] Gildas tells in a vivid passage of swarms of black curachs, crammed with hirsute, half-naked warriors, bearing down on the British coastlands.[10] The province held out the prospect of abundant loot. Flocks and herds; cloth and hides; corn, wine, and oil; pottery, glassware, and silver; coin and jewels—all this within easy striking distance of the Irish shores. Like the Vikings, the marauders had many generations of seafaring experience behind them. For well over a century, and in ever-increasing numbers, they had been coming over in search of plunder.

In these descents the Scotic curach played a part by no means dissimilar to that played by the Viking *langskip* several centuries later. Like the *langskip*, the curach was a swift and handy craft, admirably adapted to this kind of roving warfare: it could land in almost any cove, or on any sheltered strand, under conditions where it would be out of the question for heavier, larger craft to do so. Gildas, writing in the middle of the sixth century, records that the marauding *Scotti* came swarming ashore from their craft—*emergunt certatim de curucis.* It is apparent from the context that it was in large sailing-curachs that the *Scotti* arrived and departed again, year after year, with their plunder; for he relates that these craft were wafted both by the strength of oarsmen and the force of the wind.[11] It was in these curachs that much of the loot, like the Coleraine hoard, must have been carried across the seas.

What were the actual numbers of ships and fighting-men engaged in these incursions? The question can only be answered within the widest limits. It would seem that, from time to time, very large fleets indeed went forth from Ireland; and this—even allowing a good deal for poetic license—is corroborated by the testimony of Claudian. St. Patrick refers to large numbers of prisoners carried off by the Irish in the great raid in which he himself was taken captive.[12] All things considered, a reasonable assumption might be that, on the occasion of the biggest raids, fleets of several dozen curachs, crammed with warriors, were probably involved.

The part played by Irish seafarers in this era was not entirely destructive. The general conversion of Ireland to Christianity followed, after all, as a direct consequence of the abduction of St. Patrick. The increasing traffic on the western sea-routes at the close of the Imperial age materially assisted the propagation of Latin influences. The Irish curach which had for so long carried sword and fire to the coastlands of Roman Britain was destined, in the early medieva era, to bear the light of the Faith to the distant heathen lands of the west and north. In the end Irish monks and hermits carried the religion and letters of Rome to far, unknown regions where the Eagles had never penetrated. They set up the Christian symbols and recited the Christian prayers in the furthermost isles of the Hebrides. During the eighth century, at the latest, the Latin tongue was heard amid the mists and loneliness of the Faeroe Islands and in the hermit settlements along the south-east coast of Iceland.

By the first half of the fifth century the sailing-curach in use among the Irish appears to have reached a remarkably high standard of efficiency and performance. During the ensuing centuries it is heard of in many different parts of Britain, Ireland, and Brittany; it is recorded as far north as the Orkney Islands, and as far south as the vicinity of Dol, on the Brittany coast.[13] The large curach, with its single mast and square-sail, must have been a familiar sight in this era in the Irish Sea and St. George's Channel, and all along the Atlantic shores of Scotland and Ireland from the Pentland Firth to the estuary of the Shannon, from the brochs of the Orkneys and the *machair* of Iona to the rugged cliffs of Tory and the ancient duns of Aran. From time immemorial the hide-covered craft had been an integral part of the sea-communcations of the west: even as late as the ninth century it is likely that there were still some medium-sized curachs at sea.

Curachs were of different sizes, ranging from the small one-hide curach mentioned in the *Tripartite Life of St. Patrick* (in Irish, *curach oen seiched*; in Latin, *navis unius pellis*) to the large sea-going curach so often used in the Atlantic pilgrimages of Cormac ua Liatháin. The latter type is of particular interest, seeing that this kind of craft—one of the most significant products of the ancient civilization of the Gael—has long since vanished from the seas.

It is, perhaps, the fact that we can see with our eyes and touch with our finger an actual specimen of the craft of the Viking age that enables us to accept so readily the impressive achievements of Eirík Raudi and his contemporaries. It is to be regretted that Celtic archaeology can produce nothing comparable with the Oseberg, Gokstad, or Skuldelev vessels. No vestige of the large sailing-curach of the early medieval era has survived the centuries. Because of the total lack of this kind of archaeological evidence, it is impossible to trace, stage by stage, the development of the sailing-curach—as may certainly be done in the case of the Viking vessels. In the absence of this evidence it can only be presumed that the structure of the curach's hull, and also of her rigging, sail, and steering gear, were the product of a gradual, and probably very slow, evolution.

From the constructional details set out in Adomnán's *Vita S. Columbae* it is possible to visualize the sailing-curach used by the monks of Iona and the warriors of Dál Riata as a hide-covered craft of fairly substantial dimensions, propelled by a large square-sail hoisted to a mast stepped amidships. Mention is made of sail-yards (*antemnae*) and rigging (*rudentes*). Though oars were often used, the sail was the principal means of propulsion. There are numerous references to 'fair winds' and 'full sails'; it would seem likely that crews could not row against the wind. Of particular interest is the reference made by Adomnán to the vessel's keel; for the crucial importance of the keel in shipbuilding can scarcely be overstressed. Without a proper keel a craft of any size must of necessity be structurally weak and unstable; moreover, it can never be a true sailing ship. The evolution of the keel, and the development of the mast and sail, in fact, go hand in hand. It is apparent from the *Vita* and other sources that the large sailing-curach of the Age of the Saints was fitted with a proper keel and was a true sailing ship.[14]

In the *Vita* reference is made to St. Columba and his seamen endeavouring to empty the 'bilge-hole', or baling-well, which seems to suggest a craft of some size.[15] The curach was apparently still used, as in the late Bronze Age, for the carriage of merchandise.[16] On the other hand, it would seem that the curach was unable to carry substantial cargoes like the Norse *hafskip*: for once when St. Columba's followers were engaged in fetching large timbers from the mainland, instead of loading these into their curachs, they had to tow the timbers astern.[17] In both the *Vita S. Columbae* and the immrama, such as the *Immram curaig Maíle Dúin*, there is always a clear differentiation between the sailing-curach and the small rowing-curach.[18]

In the *Navigatio Sancti Brendani Abbatis*, which probably dates back to the second half of the ninth century, there is an interesting account of the construction of the medieval curach. We are told how Brendan and his companions fashioned a very light craft, with ribs and sides of wickerwork, 'as is the custom in that country', and covered it with ox-hides which had been tanned with oak-bark, caulking the joints with pitch. The sail was hoisted to a mast stepped amidships. Provision was also made for steering the craft (though no details are given of this). The vessel carried a crew of seventeen and provisions for forty days, in addition to a supply of butter for greasing the ox-hides, should the curach's skin-covering require repair.[19]

Certain items of a curach's gear are worth noticing. In the *Immram curaig Maíle Dúin* mention is made of 'oars' and 'rudder', *mo rama ocus mo lui*.[20] This suggests that some kind of separate steering device was in use and not simply one of the oars. In an early *Life of St. Brendan* (*Vita Prima*) there is an interesting reference to what was evidently an iron anchor. That it really was an iron anchor is shown by the remark made by St. Brendan to one of his crew to do 'the work of a smith'.[21] This anchor may have come to Ireland either from the North, in the Viking era; or else, at an earlier period, from some part of the Roman Empire. It will be recalled how in Caesar's *De bello gallico* mention is made of the *ancora*, or iron anchor, on the Armorican coast.[22]

The provisioning of these early Irish craft presents certain features of interest. In the *Vita S. Columbae* mention is made of a leathern milk-bag; in the *Betha Brenainn Cluana Ferta* it is related that stores of herbs and seeds were taken on board St. Brendan's curach. In both the *Vita Prima* of St. Brendan and the *Immram curaig Maíle Dúin* reference is made to 'fish' and 'salted fish'. In the former work wheaten bread, herbs, water, and fruit are likewise mentioned. It is possible and even probable that these Irish, like the Norsemen, partook of a rather more wholesome diet on shipboard than did mariners of a later era. In this connection it is to be noted that, in the time of the Norse emigration to Iceland, a band of Irish thralls are recorded to have imparted a useful piece of sea-lore in respect of diet to their Scandinavian shipmates. When it was intended to establish a new settlement, in addition to stores for the voyage the colonists presumably carried the necessary livestock along with them. As will later be seen, it was apparently Irish settlers who first stocked the Faeroe Islands with sheep.[23]

There are three little-known references to the early medieval sailing-curach which may be quoted here. In the story of 'Branwen daughter of Llyr', in the Welsh *Mabinogion*, the following passage occurs, from which it would appear that curachs still crossed the Irish Sea, as in the days of St. Patrick and Sidonius. '"Aye, Lord," said his men to Matholwch, "set now a ban on the ships and the ferry-boats and the coracles [*corygeu*], and such as come hither from Wales, so that none may go [from Ireland] to Wales; imprison them and let them not go back.'"[24] It would appear that the craft used by the warriors of Dál Riata on their forays were sea-going curachs; for it is recorded that on one of these ventures, in 622, Conaing, son of Aedán mac Gabráin, met his end when his curach was lost at sea.[25] In the Annals of Ulster, in 641, an entry occurs relating to the loss of a vessel belonging to the monks of Iona which may refer to the hide-covered curach, 'naufragium scaphae familiae Iae'.[26]

Nor is there any sign or suggestion in the *Vita S. Columbae*, which was compiled late in the seventh century, that the curach was then obsolete. It is even possible that, as late as 828, when Diarmaid, Abbot of Iona, crossed from Ireland to Scotland with the relics of St. Columba, he voyaged thither in a curach.[27] After Adomnán's death the voyages of Irish clerics are seldom mentioned in detail, and it is practically impossible to discover what type of vessel was used on any particular venture. The details which the learnèd geographer Dicuil (one of the Irish scholars residing at the court of Charlemagne) gives of a voyage to the Faeroe Islands in 825 might equally apply to a curach or to a planked vessel.

Roughly speaking, the curach era may be said to have terminated with the advent of the Viking war-fleets to Ireland towards the end of the eighth century. *dating* The comparatively peaceful conditions under which the overseas enterprise of the Irish clergy had flourished were brought abruptly to an end. The anchorites abandoned their settlements in the Faeroe Islands, and, several generations later, the Iceland voyage. The Norse influence upon the maritime activities of the Irish during this period is significantly reflected in the long list of Norse loan-words in Irish relating to seafaring and trade. With the exception of *curach* and *ram* and a few other words almost the entire vocabulary of shipping and commerce was henceforth of foreign origin.[28] Still later, as the twelfth-century *Vita Niniani* suggests, all that was left of the historic Irish sailing-curach in Ireland and Scotland was the memory.[29]

Almost all the references to hide-covered craft that still occur evidently allude to the small rowing-curach, or to the coracle. The hide-covered craft referred to in early Welsh literature and in popular sayings—with the single exception of the sea-going *corygeu* in the *Mabinogion* cited above— are, one and all, coracles. Gerald of Wales, about 1200, tells of a hide-covered curach of 'moderate size', *cymbulam modicam*, which (he says) carried two men and was propelled by oars.[30] But this was clearly not a sailing-curach of the old type. Nor, by all accounts, was the *curachan* in use in western Scotland and the Isles as late as the eighteenth century anything more than a small rowing-curach or coracle.[31] 'In size if not in constructional features these craft appear to have degenerated', commented the late James Hornell, who carried out a comprehensive survey of the modern curach.[32]

With Roderick O'Flaherty's *Ogygia*, published in 1685, begins the academic study of the curach and its history.[33] It is, perhaps, worth noticing that this study dates back to an era when the sailing-curach may still have survived on some of the inland waters of Ireland. Both Froissart and Holinshed mention large coracles capable of carrying troops during the Hundred Years War.[34] That, much later than this, the tradition of constructing wickerwork vessels of substantial size had never completely died out may be surmised from the fact that during the Irish wars at the close of Elizabeth's reign a band of fugitives under O'Sullivan Beare escaped across the Shannon (the enemy having removed all skiffs and boats) in an improvised craft made of light timber and osiers covered with the hides of a number of horses they had slaughtered. O'Sullivan Beare's followers actually built two vessels: one fairly large, of eleven hides, and one quite small, of only one hide. These vessels were built according to the traditional method—i.e. bottom up. Reference is made to strong wooden gunnels, *solida tabula statumina*, and thwarts, for the rowers; thole-pins also were fitted. The vessels were borne by night on the soldiers' backs from Bosnach wood to the river at Port-a-tulchain. The smaller craft was found to be useless; but the larger, which would take thirty men at a time, carried the whole band across the water in safety.[35]

Finally, in the Pepysian Library at Magdalene College, Cambridge, there is an interesting sketch of what is apparently some kind of sailing-curach, which is considerably larger than the modern rowing-curach. It was made by Captain Thomas Phillips, 'Seaman and One of His Majesty's Tower Engineers', at some time during the reign of James II. The sketch is in two divisions, the upper of which—inscribed a *Portable Vessel of Wicker, ordinarily used by the Wild Irish*— depicts a sailing-curach as large as a Galway hooker, bluff-bowed, broad-beamed, and propelled by a square-sail (many times bigger than that commonly used in the modern curach) hoisted to a mast stepped amidships. The mast is secured by two stays on each side. The sail-yard is raised or lowered by a halyard made fast in the bow. In the background may be seen two other sailing-curachs. The lower of the two divisions is inscribed: '*The Method of Workeing up ye s*^{d} *Vessel us'd by ye Wilde Irish; taken upon the place by Capt. Tho. Phillips.*' The bow, keel, and gunnel are of substantial scantlings, while the ribs between the keel and gunnel are formed by withies, covered over with a skin of either canvas or hides. The sketch is in certain respects almost certainly inaccurate; and it is scarcely surprising that its authenticity has been called in question. However, in the volume of manuscripts of which this drawing forms part there are several other folios based on materials supplied by Captain Phillips, who appears to have been an experienced observer of ship-types that interested him. Even if the drawing cannot be accepted as an altogether accurate representation of a contemporary sailing-curach, therefore, it may very well have been a crude sketch of some vessel which Captain Phillips had either seen on his travels or had had described to him.[36]

The curach of the present day is primarily an oared craft, the small sail serving merely as an auxiliary. The hull is formed of a framework of thin wooden lathes or withies, covered over with stout tarred calico (the ancient cow-hide covering went out about the middle of the last century). It is light enough to be carried, bottom up, on the shoulders of its crew. It is capable of taking a load of about two tons. It is easy to manœuvre and rides the long western seas as lightly as a gull.

Such are the buoyancy and handiness of the craft that it is said to be virtually unsinkable so long as the courage and strength of the crew hold out. Centuries of experience have bred a most marvellous skill. Good curachmen seem to be almost part of the craft itself. On the very infrequent occasions when one of them is lost, it usually transpires that the crew 'had drink taken'. Three first-rate men in a good curach can outlive the worst gales, as experience has often shown. The curach can take a breaker and swim even when three-quarters full. One has been told of curachs remaining afloat under conditions of weather and sea in which even the local lifeboat dared not put out. In this connection a tribute to the seaworthiness of a curach, from Donegal in the mid-nineteenth century, may well be quoted. 'An old cutter's man . . . agreed in considering them to be the safest of all boats in the hands of men accustomed to their management; during all his experience in the Sound of Tory, he never knew or heard of one being lost, though they ventured out in all weathers. . . . Cattle are transported across the Sound in these boats.'[37] Another tribute may be mentioned in reference to the curach in the Aran Islands. 'Such is the dexterity with which it is usually manœuvred, that it will land seamen from ships in distress through the roughest breakers, and cross over to the mainland, when vessels of every other class are unserviceable.'[37]

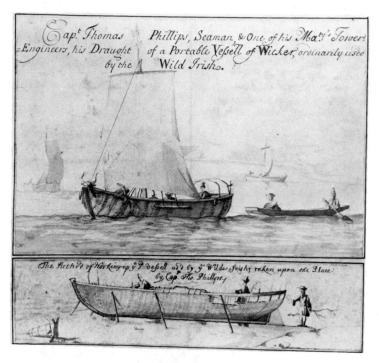

'A Portable Vessel of Wicker'

The curach is admirably adapted to the conditions under which it is used on the Atlantic seaboard of Ireland. Along the irregular, broken shores of Donegal, Sligo, Mayo, Connemara, Clare, Kerry, and Cork, fringed by so many isles, rocks, shoals, and outlying dangers, experience has shown that it is possible to get a curach in and out, on many parts of the coast where there is neither harbour nor slipway, in almost any kind of weather. Even on an exposed rocky strand a curach is sometimes snatched bodily out of the breakers and run to safety. (In bad weather this would be occasionally done at the small fishing village of Gort na gCapall, on the iron-bound southern shore of Inishmore, Aran Islands, and in similar situations.) Such a craft is well fitted to skim between the islands and run in and out of the intricate channels of the long, deeply indented Atlantic coastline.

The dexterity of an experienced crew in a seaway is amazing to behold. There is a description of such a scene in J. M. Synge's *The Aran Islands* which is worth quoting here.

Every instant the steersman whirled us round with a sudden stroke of his oar, the prow reared up and then fell into the next furrow with a crash, throwing up masses of spray. As it did so, the stern in its turn was thrown up, and both the steersman, who let go his oar and clung with both hands to the gunnel, and myself, were lifted high above the sea.

The wave passed, we regained our course and rowed violently for a few yards, when the same manœuvre had to be repeated. As we worked out into the sound we began to meet another class of waves, that could be seen for some distance towering above the rest.

When one of these came in sight, the first effort was to get beyond its reach. The steersman began crying out in Gaelic, *Siubhal, siubhal* ('Run, run'), and sometimes, when the mass was gliding towards us with horrible speed, his voice rose to a shriek. Then the rowers themselves took up the cry, and the curagh seemed to leap and quiver with the frantic terror of a beast till the wave passed behind it or fell with a crash beside the stern.

It was in this racing with the waves that our chief danger lay. If the wave could be avoided, it was better to do so; but if it overtook us while we were trying to escape, and caught us on the broadside, our destruction was certain. I could see the steersman quivering with the excitement of his task, for any error in his judgment would have swamped us.[38]

In bad weather the water would be coming aboard all the time 'in pints', and the middle man in the curach would be kept feverishly baling. If a lot of water came aboard suddenly, they must bale out before the next sea swamped them. Under such conditions there was danger, not merely of being swamped, but of a big sea actually capsizing the curach. The bowman was continually on the look-out for such a sea, in readiness to turn the prow of the curach with a few strong pulls. If a bad sea could not be avoided it had to be rushed, before the oncoming comber

Launching a Curach, Aran Islands, c. 1920

should curl over too far. A cross-sea, occasioned by a sudden change of the wind, was especially dangerous; even so, a curach would be manœuvred with wonderful skill through the turmoil of clashing seas.

Many was the time, in bygone days, when the Aranmen were miles out in the Atlantic long-lining, and, the weather having turned suddenly bad, crowds of old men and women would be up on the crags anxiously watching for the return of the curachs. With the wind and sea making fast, the crews would have a stiff fight to get in from 'out at the back' and bring their craft safely to land. Once, long ago, a four-oared curach was caught in this way when fishing 'out at the back' . . . only her oars ever came ashore.

An epic feat of seamanship is remembered to this day in the Aran Islands. A three-oared curach, long-lining some six miles off the entrance to Gregory Sound, was caught by a southerly gale. Her crew dared not run before the heavy seas for fear of being swamped; they could only hold her head on to the sea, and, whenever they got the chance, bring her back on her proper course for a short, swift run, before pulling her round once again to face the seas. She was driven down Gregory Sound past the low cliffs behind the Glassan Rocks thronged with islanders watching the curach fighting for her life. In vain the crew endeavoured to reach the strand from which they had set out early that morning; but wind and sea were too much for them. The curach was driven right across Galway Bay; and, finally, with her bow still firmly held to the sea and wind, came to land at Salthill, near Galway town.[39]

4

Early ventures in the Atlantic

WONDERFUL AND IMPRESSIVE as are the performances of some of the Irish curachmen of the present day, however, they do not for a moment compare with the achievements of their forbears in the Age of the Saints. Urged on by the desire to discover some sea-girt refuge where they might work out their salvation, remote from the troubles and temptations of the world, the Irish hermits voyaged to the Hebrides, to the Orkneys and Shetlands, and at last to the Faeroes, where they appear to have settled about the year 700 and remained for the next hundred years, till driven out by the marauding Norsemen.[1] Impelled by very different motives from those which inspired the latter, they nevertheless preceded the Vikings in some of their most daring ventures.

It is difficult to say when and how these clerical ventures into the Western Ocean began. Though St. Brendan was one of the earliest, and certainly by far the best known of the voyagers, there is no reason to suppose that he was the first. It may well be that even before Gildas wrote the *De excidio*, at a time when there were still living in this island those who could recall the thronged and busy *fora* of the towns of Roman Britain, and the agreeable life of the great country houses, in the last phase of the Imperial era, some intrepid Irish mariner was steering his curach out into the open sea.

In this era, perhaps the most brilliant in her history, Ireland was not only a centre of learning renowned throughout Europe, but sent forth her monks over the seas to France, Germany, Switzerland, and even as far southward as the plains of Lombardy. 'To voyage over seas and to pass over broad tracts of land', the Briton, Gildas, testified of the Irish, 'was to them not so much weariness as a delight.' The Atlantic journeyings of the Irish monks and anchorites recorded by Adomnán represented only one facet of all this intense activity. Between their missionary enterprises and certain of the great monastic schools—those of Clonfert and Bangor in particular—there was a close and vital connection. Further, the Christian clerics were the heirs of an ancient seafaring tradition. They sailed the seas like their heathen ancestors before them, and in the end travelled further than any chieftain in the history of the Gael.

As has already been said, the outstanding figure among these Christian seafarers of the Age of the Saints may be said to have been St. Brendan of Clonfert, who, in the early half of the sixth century, was voyaging at least to the Hebrides.[2] Of all the *immrama*, certainly his was, and is, the most widely read. There are good reasons for rejecting the belief that all the discoveries attributed to St. Brendan were actually made by the Saint in question. Rather they may be said to represent the collective sea-experience of successive generations of Irish mariners; and regarded in this light, it is more than possible that certain of the adventures described in the Latin and Irish texts were founded upon fact. What is, however, perfectly certain is that the name of St. Brendan was writ large upon the coasts and islands of the west. According to an ancient Faeroese tradition the Saint and his followers landed on Stromo Island, in Brandansvík, or Brendan's Bay. The St. Brendan of history at any rate voyaged widely up and down the Atlantic coasts of Ireland and Scotland. In the Hebrides, the church in Barra and the monastery in Tiree were dedicated to St. Brendan, whose name was also preserved in Kilbrandan Sound, separating Arran from Kintyre. He is associated as well with Lorne, Mull, Islay, the Garvellach Islands,[3] and Bute. Further south, St. Brendan is commemorated in his native Kerry in Brandon Hill, Peak, Point, and Bay (where he is said to have constructed his famous curach), and in Brandon Rock in Mal Bay, Co. Clare. The mythical 'St. Brendan's Isle' was believed in by mariners throughout the Middle Ages, and the no less mythical 'Brendan's Rock' did not finally disappear from British Admiralty charts until well on in the nineteenth century.

In an interesting and revealing article W. F. Thrall drew attention to the significant fact that the *Immrama* 'all seem to have sixth-century settings. The sixth century seems also to be the age of the first great vogue of the sea pilgrim'. And he quoted the following entries, 'all seemingly connected with the sixth-century saints', in the ancient *Litany of Oengus*.

Thrice fifty true pilgrims who went with Buti beyond the sea.
The twelve pilgrims who went beyond the sea with Moedhog of Ferns.
Twelve men who went beyond the sea with Rioc, son of Loega.
Thrice twenty men who went with Brendan to seek the land of promise.
The twelve youths of whom Brendan found the survivor in the Island of the Cat.
Three descendants of Corra, with their seven companions.
Twelve men who encountered death with Ailbe.
Four-and-twenty from Munster who went with Ailbe upon the sea to find the land in which
    Christians ever dwell.
Twelve youths who went to heaven with Molaise without sickness.
The confessor whom Brendan met in the promised land, with all the saints who have
    perished in the isles of the ocean.[4]

'The travels of the Celtic Saints at that epoch', observes Frere, 'match those of the megalithic missionaries over the same waters three thousand years earlier.'[5] How far they went and what they achieved will in all probability never be known. It is impossible, across the abyss of time, to distinguish between fact and legend. But the signs of a strong surge of maritime activity are unmistakable.

Of the voyages of Columba and his followers we have far more information than of these earlier sea-pilgrims. Adomnán wrote that Columba's name was known throughout all the regions of the isles of the ocean, 'cuius nomen Columba per omnes insularum oceani provincias divulgabitur notum'.[6] In the course of the seventh and eighth centuries the monks of the *familia Columbae* and other communities settled themselves in a great many of the isles and islets off the Irish and Scottish coasts. Even at the present day the remains of ancient ecclesiastical settlements are still to be seen in Tory Island (Co. Donegal), Inishmurray (Co. Sligo), Inishglora (Co. Mayo), the Aran Islands (Co. Galway), and the almost inaccessible Skellig Michael, off the west coast of Kerry. The monks voyaged

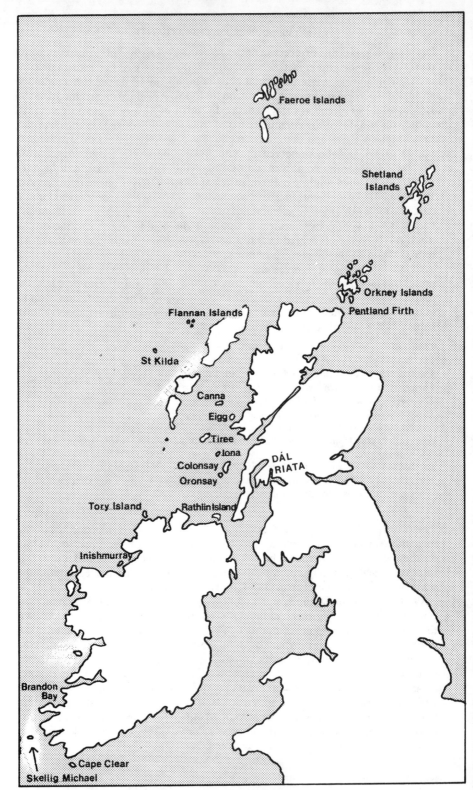

Irish maritime enterprise in the Age of the Saints

further and further north, to the Garvellach Islands, Oronsay, Colonsay, Tiree, Ardnamurchan, Lochaber, Eigg, Canna, and Skye. The remains of their small chapels are to be found in such remote localities as the Flannan Islands and North Rona. The work of St. Columba and his monks opened the north of Britain to the spiritual and cultural influences of Irish Christianity and marked one of the great turning-points in Scottish history.[7]

The close links of the community with Ireland were maintained. For hundreds of years to come Iona remained virtually an Irish island. At the same time it was the spiritual capital of Scotland. When the king of Dál Riata died, it was St. Columba, according to Adomnán, who appointed and consecrated his successor, Aedán mac Gabráin. It was to Martyrs' Bay, in Iona, that galleys bore the royal dead for several centuries. In the seventh and eighth centuries the spiritual empire of the Irish monks reached its culmination.

From scattered fragments of verse and marginal jottings in early manuscripts is revealed something of the intense interest and delight some of these monks took in the wild life about them, in the woods, the shore, the sea, and the ever-changing skies, and in all the loveliness of nature. From some such sea-girt hermitage as Skellig Michael, they would gaze down on the groundswell fretting on the rocks below, the gannets from Little Skellig diving after fish, and the black hags skimming low over the waves, and look upwards at the white scarves of mist clinging perpetually to the twin pinnacles of their rocky islet. As a background to their devotions they would hear the plaintive wailing of the herring-gulls wheeling and circling about the crags, the shrill cry of the oyster-catcher, 'the bird of St. Bride', and, day in, day out, in the hours of light and in the hours of darkness, the low sullen thunder of the groundsea.

No less than four ocean pilgrimages are recorded by Adomnán in the *Vita*. Three of them were made by the monk Cormac ua Liatháin, 'a truly pious man', who went in search of a 'desert' (that is, a 'desert island') in the ocean, but did not find it. Adomnán states that on the second of these ventures Cormac set out from Eirros Domno (Erris in Co. Mayo); and that, in the course of what was evidently a long voyage, he visited the Orcades, or Orkney Islands. He was away for several months and eventually returned to Iona. Cormac's third voyage, from the navigational standpoint, is by far the most interesting and important. According to Adomnán, for fourteen summer days and nights his curach sped with full sails before the south wind in a straight course from land, until his voyage seemed to be 'extended beyond the limit of human wanderings, and return to be impossible'. No mention is made of the point of departure, and all that is known about Cormac's third voyage is that his course was northerly, and what in the end obliged him to turn back seems to have been an alarming encounter with a shoal of red jelly-fish, or similar species.[8] What is particularly important about this narration is that it proves beyond question that the voyage was made in a sailing-curach. In it Adomnán refers to the keel, prow, stern, and leathern skin, also the mast, sailyard, sail, and rigging of the craft used by Cormac ua Liatháin. In all probability the other Atlantic voyages mentioned by Adomnán were made in curachs also: since it was only because of the misadventure which obliged them to turn back that the occasion arose for describing the craft in which they sailed. It is significant, too, that in the *immrama*, such as the *Immram curaig Maíle Dúin*, the craft which is almost invariably used for these ocean pilgrimages is a curach.[9]

Cormac's ventures into the Western Ocean must have made a considerable impression upon his contemporaries. Though Adomnán is not especially interested in ocean navigation for its own sake, he has recorded each of these voyages in the *Vita S. Columbae*; and he also mentions one Baitán, who before setting out asked a blessing of Columba and then with others 'sailed in search of a desert in the ocean'. The same Baitán, we are told, after long wanderings in stormy seas finally returned to his native land without finding the 'desert'.[10]

Skellig Michael

Apart from the missionary and eremitical ventures of the Irish clergy, there was a good deal of activity on the western sea-routes in the early Middle Ages. Sea-links which for centuries had languished while land-communications were secure under the sway of the *pax Romana* regained their importance when the Saxons assailed and gradually overran the lowland zone of Britain. By the sixth century, indeed, traffic was plying between these western lands and the Mediterranean.[11]

Moreover, without adequate sea-transport it would have been impossible for the kings and chieftains of Ireland to engage in far-ranging wars and forays against their neighbours to the east and north. So numerous were the hide-covered craft used in one of these incursions, according to Irish tradition, that the waters between Ireland and Scotland were spanned by a 'bridge of curachs'.[12] These operations must have served as fertile breeding-ground of experienced mariners.

The Irish Sea was virtually an Irish lake. In the late Imperial era the Irish had established a number of colonies up and down the western coastlands of Britain; the most extensive of which was Dál Riata, in the west of Scotland. Dál Riata was an offshoot from the Irish kingdom of Dál Riata. The importance of the sea to Dál Riata, with its difficult and scanty land-communications, will be readily comprehended. It was the sea, after all, which bound together and unified the various parts of the kingdom. Throughout the early medieval era there was a continual coming and going between Ireland and the Scottish Dál Riata.[13]

In the maritime enterprise of the Irish, as in that of the Norsemen, geography was a determining factor. Its influence can scarcely be set too high. The Atlantic shores of Ireland and Scotland are in certain significant respects not two coasts, but one coast. As W. R. Kermack has rightly observed, 'There stretch out from Scotland like a bridge between the two countries, the peninsula of Kintyre and the island chain of Jura and Islay, so nearly reaching across that the Mull of Kintyre is distant only thirteen miles from Fair Head in Antrim. The coastlines both of western Scotland and southern Antrim are fretted into bays and fjords such as formed the nursery of the Vikings; such coasts are brought together, rather than separated by salt water, if either coast is inhabited by a people apt to the sea.'[14] Such conditions did, in fact, then exist; and at the same time it is to be remembered that the Atlantic littoral of Ireland, like that of Scotland, was emphatically a 'curach coast', where these vessels could be readily run ashore in some sheltered bay or cove and hauled up above the high-water mark.[15]

When Aedán mac Gabráin became king in 574 Dál Riata was a strong and growing power in the north. Under this king Dál Riata began to expand to the eastward. Aedán mac Gabráin, the contemporary and patron of St. Columba, was a strong and able ruler. It is recorded that in 579 he led an expedition against the Orkney Islands. (A story which has been lost, *Eachtra Aedain mac Gabrain*, is believed to have been concerned with this expedition.) The object of this campaign is unknown.[16]

Dál Riata, a country of strong maritime traditions, was organized for the manning of large war-fleets capable of undertaking far-ranging expeditions. Every household was required to furnish a given quota of oarsmen. In the *Senchus Fer nAlban*, the 'History of the men of Scotland', ascribed to the seventh century, mention is made of the system of recruitment for the considerable war-fleet maintained by Dál Riata. Throughout the seventh century, and, indeed, until well on in the eighth, it is apparent from numerous entries in the *Annals of Ulster*, the *Chronicle of Iona*, and other sources that the Irish of Dál Riata remained a warlike, seafaring people, who were able to raise ships enough and seamen enough to launch successive expeditions by sea against their enemies.[17]

The Pictish hegemony of Scotland may be traced back to the middle of the fourth century. Both Picts and Irish became formidable foes of Roman Britain in the late Imperial era. Little is known for certain about the Picts. Those of them who lived on the islands of the north and west are supposed to have been, for the most part, mariners; and those who dwelt on the mainland, chiefly farmers. It is possible that they used hide-covered sailing craft like those of the Irish and of the British Celts of the Cornish peninsula; but there is no certain evidence on this point. In most cases there is no satisfactory method of distinguishing between the part played by the Picts in these descents on Roman Britain and that played by the Irish. In St.

Columba's time the main centre of the Pictish power lay apparently north of the
Tay. The Picts also held the Northern Isles. From the sixth century onward
intermittent warfare was carried on between the Picts and the Scottish Dál Riata.
Aedán mac Gabráin was engaged in fighting the Picts for many years, and with
varying fortune. This intermittent warfare continued under Aedán mac Gabráin's
successors.[18]

Both Picts and Irish were vigorous, bellicose peoples possessed of numerous war-
craft.[19] The territorial power and influence of both were upheld and extended with
the aid of fleets. But the Pictish achievement at sea could scarcely be compared with
the Irish. The enterprises of the former were limited to what might be termed home
waters. The Picts could boast of no such record of deep-sea navigation or daring
exploration as that of the Irish. To say, as Professor A. R. Lewis has said in *The
Northern Seas*, that 'this Pictish population had by the seventh century again
developed a maritime tradition quite as important as the Irish'[20] is certainly an
overstatement; and this same authority's belief that the Picts were sailing to Iceland
is pure fantasy. The claim to be the pioneers of ocean navigation in northern waters
belongs, without any question, to the Irish.

The arrival of the Irish missionaries in the Orkney Islands can be assigned, with
reasonable confidence, to the latter half of the sixth century. Already about the year
565 St. Columba had used his influence with the Pictish king, Brude, to ensure that
Cormac ua Liatháin should be duly protected if he happened to land in those
islands.[21] Later the monks established permanent settlements there. The memory of
these anchorites is preserved in a number of significant place-names in both
Orkneys and Shetlands: for example, Papley, Papa Westray, Papa Stronsay, Papil
Stour, and Papil Ness. The name 'Munkerhoose' is found in Papa Westray and also
in Auskerry, while in Stromness there is 'Monker House' and in Eynhallow
'Monkerness'. Fragments of early Christian sculptures have been discovered in Papa
Stronsay in the Orkneys and at Papil in West Burra in the Shetlands. There are also
suggestive ecclesiastical remains—including an early Celtic church bell—on the
tidal isle known as the Brough of Birsay.[22]

On some such venture as Cormac's the Irish may have eventually discovered the
far-off Faeroes. Whether or not the discovery occurred during the first great age of
the sea-pilgrimage, the sixth century, or at some later period, is unknown.
According to Dicuil, his countrymen had settled in distant islands to the northward
of Britain (which were presumably the Faeroes) about the beginning of the eighth
century. Later they had fled before the Norsemen, and the islands were now
uninhabited. Dicuil's account is as usual clear and precise, and seems to have been
based on good authority.

There are many more islands in the ocean north of Britain, which can be reached from the
northern British Isles in two days' and two nights' direct sailing with a full sail and a
favourable wind. A trustworthy priest (*presbyter religiosus*) told me that he had sailed for two
summer days and an intervening night in a little boat with two thwarts, and landed on one
of these islands. These islands are for the most part small; nearly all are divided from one
another by narrow sounds, and upon them anchorites, who proceeded from our Scotia
[Ireland] have lived for about a hundred years (*in centum ferme annis*). But as since the
beginning of the world they had always been deserted, so are they now by reason of the
Northern pirates emptied of anchorites, but full of innumerable sheep and a great number of

different kinds of seabirds. We have never found these islands spoken of in the books of authors [i.e. classical authorities].[23]

By about the year 700 the Irish hermits had settled in the islands: that much is clear. They dwelt there for about a century. The flocks which they had brought with them increased and multiplied. Traces of the ancient Celtic field system, notwithstanding the intervening centuries of Norse occupation, remain to this day. There was formerly a *papar* place-name in the Faeroes: Papurshalsur.[24] But the date of the Irish discovery of those distant isles, 'deserted since the beginning of the world', and unknown to any of the geographers of antiquity (as Dicuil declared), is shrouded in mystery. Whenever it actually occurred, it was an outstanding achievement—the first notable geographical discovery of the Middle Ages.[25]

5

The first discovery of Iceland

IN THE OPINION of various modern authorities, it may be argued that Iceland's story starts in a remoter past even than in the early Middle Ages, when it was discovered, first, by the Irish and, later, by the Norsemen. In a number of rather obscure fragments attributed to the Greek Pytheas, mention was made of 'the island of Thule', supposed to be situated about six days' sailing to the north of Britain, and one day's sailing from 'the frozen sea'; around the island, it was further related, there was neither sea nor air, but 'a mixture like a sea-lung'. Reference was also made to certain barbarians who inhabited Thule.

When all is said and done, however, there is really no evidence worthy of the name to substantiate the theory that Iceland was known to the ancient Greeks. Pytheas, when he speaks of 'Thule', assuredly does not mean Iceland, but possibly—as Bugge has conjectured—one of the inhabited islands off the west coast of Norway, where the astronomical conditions are similar to those described by the Greek geographer.[1]

Another and much more recent theory is that Iceland was visited, on at least one occasion, by a vessel from some part of the Roman Empire. The evidence adduced for this supposition is almost entirely archaeological. Within the last sixty years three small copper coins of the Imperial age—Aurelian (270–5), Probus (276–82), and Diocletian (284–305)—have been discovered in or about the farm of Bragdhavellir, on the south-east coast of Iceland. The wholly unexpected presence of such relics on the shores of an island lying several hundreds of miles out in the Atlantic has naturally given rise to a number of ingenious conjectures.

In an able and most interesting paper by Dr. Kristján Eldjárn it was suggested that these coins were brought to Iceland by some Roman seafarers at a time when they were still current; namely, in the later Imperial age. Eldjárn based his argument on the extreme rarity of Roman copper coins of this era outside the limits of the Empire; for the same reason he rejected the possibility of the coins having been carried to Iceland by the *papar*; and he enlarged on the 'colossal activity' at sea in the time of Carausius; concluding that in the days of Carausius' rule in Britain (287–96) a Romano–British vessel was driven off its course and eventually carried by the current to that part of the Icelandic coast which was to witness so many wrecks and strandings in the centuries to come. A tradition about this voyage, wrote Eldjárn, might have lived on in the British Isles and in the course of time given the Irish anchorites, and later the Norsemen, the idea of seeking a country in the North Atlantic.[2]

'Colossal activity' is, perhaps, an exaggeration. The *classis Britannica* was possibly a crucial factor in the chain of causation which led up to Carausius's revolt; and it is true that Constantius eventually restored the Imperial authority in a series of highly successful—and presumably well-organized—combined operations against the rebel forces. But there are no grounds for supposing that the Roman war-galleys operated off any northern part of our islands—for example, the Orkneys or Shetlands—whence they might have been blown to Iceland. In later centuries vessels are recorded to have been driven to Iceland from the vicinity of the Faeroe Islands, and also on passage from Bergen to Scotland; but never from *southerly* regions.

The late Dr. Haakon Shetelig also apparently accepted this explanation of the presence of the three copper antoninians on Icelandic soil. His conclusion was that the Roman coins from Bragdhavellir raised the question whether the Irish were really the first discoverers of Iceland, or whether they might perhaps have been acquainted with ancient reports, which were once known to seafarers of Roman Britain, 'of the fantastic island, covered with ice and flaming mountains, far away in the northern mists of the Ocean'.[3]

There are, nevertheless, the strongest objections to this solution of the problem. The rarity of Roman copper coins of this era is certainly a powerful argument: but scarcely a conclusive one. The possibility that at some later date the coins were brought to Iceland—perhaps as children's playthings, perhaps, even, as a hoax—is by no means to be ruled out. A well-known example of this kind of fortuitous arrival is the case of the 'Prince's Flowers' on the *machair* of Eriskay: such accidents can and do happen. It is to be observed, moreover, that the most careful examination of the sites along the south-east coast of Iceland where these coins were found has failed to yield any corroborative evidence; and, above all, it is more than doubtful whether any Romano–British ships of the period were sufficiently seaworthy to make the long ocean crossing to Iceland.

In the early Middle Ages the situation is quite different. By this time the Irish had evolved a distinctive type of sailing-vessel which was admirably adapted to ocean navigation. The qualities of the curach had already been tested and proven by long experience. In the same way the early Norse traffic to the Faeroe Islands and to Iceland was made possible, a century or so later, by the evolution of the *knörr*, an ocean-going sailing ship of which more will be heard presently. Again, the discovery of Iceland by the Irish, and its eventual rediscovery by the Norsemen, formed, on both occasions, part of a natural and gradual expansion northward and westward across the Atlantic. In both cases the discovery was preceded by that of other island-groups—namely, the Orkneys, Shetlands, and Faeroe Islands. The Irish, as we have seen, were also the first discoverers of the last-named group.[4]

In short, there is no convincing evidence that man ever visited Iceland prior to the early medieval era. Both Iceland and the Faeroes were in fact unknown to the geographers of antiquity. Thule of the ancients, wherever it may have been, was certainly not Iceland.

The question now arises, at what date was Iceland discovered by the Irish? Only within very broad limits is it possible to suggest an answer. In the first place, it is hardly likely that the way there was known as early as St. Columba's time, since the third voyage made by Cormac ua Laitháin is said to have expanded 'beyond the limit of human wanderings'; and no mention is made of any land resembling

Iceland. The northerly course steered by Cormac on this voyage does suggest Iceland as a likely destination. On the other hand, the Faeroes also lay to the north; and Baitán, as well as Cormac, was searching for 'a desert in the ocean'. Whether this desert island of Baitán's lay in the same region as Cormac's, or a different one, we have no means of knowing. All things considered, judging from the fact that Cormac's long venture across the open sea was evidently regarded as something without precedent, it scarcely appears probable that communication with Iceland had been opened up as early as this.

There is also the testimony of the *immrama* to be taken into account. In these strange and bewildering narrations of legendary voyages made in curachs, it is possible now and then to catch a glimpse of various Atlantic islands and islets seen, as it were, through a glass darkly. Thus it is not improbable that Iceland, that volcanic isle, is the subject of a somewhat obscure reference in the *Immram curaig Maíle Dúin*.[5] This work, with its allusions to pilgrimages across the ocean, solitary hermits dwelling upon islands, the rowing and sailing of curachs, and, above all, in its manifest awe of 'the great, the endless ocean',[6] bears a striking resemblance, in certain regards, to the *Vita S. Columbae*. Apparently, however, it belongs to a rather later period than the *Vita*; suggesting as it does voyages and discoveries far beyond the range of those mentioned by Adomnán. The 'high, mountainous island full of birds',[7] as well as the island inhabited by huge flocks of sheep,[8] seem to suggest the Faeroe Islands, and the 'great silvern column',[9] which was apparently an iceberg, may well point to voyages to the westward even of Iceland: since icebergs are not to be met with on the eastern side of the Atlantic. It would appear that the original version of the *Immram curaig Maíle Dúin* may be assigned to the eighth century, or perhaps even earlier;[10] from which it follows that the Irish may in fact have been sailing to the island of glaciers and volcanoes[11] and settling in the Isle of Sheep (presumably the Faeroe Islands) at a date considerably anterior to that of the first recorded voyage to Iceland in 795.

How was Iceland discovered by the Irish?

A suggestion perhaps worth considering is that this venture, as well as the later enterprise of the Norsemen, might be attributed to knowledge derived, at any rate in part, from the Arctic mirage.

From time to time, under certain atmospheric conditions (particularly in summer, which in the early medieval era was the normal sailing season), the Arctic mirage almost inevitably occurs. When air rests on a significantly colder surface the observed image may be optically displaced, in a vertical direction, from its true location (this is the *hillingar* effect, or 'upheaving'), which permits the occasional sighting of objects, such as islands and mountains, situated far beyond the normal distance to the horizon.

The Arctic mirage may thus have played a part in the early Irish ventures, striking ever further out into the Atlantic, and eventually extending to the Faeroe Islands and to Iceland. It is to be observed that the most favourable conditions for this phenomenon occur from dawn to shortly after sunrise. 'The direction of potential discovery was to the north-west and the observer on or near the Faeroe Islands would have the light of the morning sun over his shoulder and in consequence would enjoy the best viewing conditions possible.'[12]

Another possible explanation is that an Irish craft making for Iceland may have followed the route of migrant birds making from the west of Scotland or the

Faeroes. One of these main migratory routes passed up the western coast of Scotland and thence to the Faeroe Islands and Iceland.[13] In the former group of islands, as we know from Dicuil, Irish anchorites had been established throughout the eighth century.

Again, an Irish vessel making for the Faeroe Islands from somewhere to the southward may have been blown off its course by an easterly gale and thus have fetched up on the south-east coast of Iceland.

That a voyage was made to Iceland in the year 795 admits of no doubt. According to Dicuil it was made by a band of Irish clerics. Judging from the testimony of his informants, it would appear that they sailed for the island at some time during January, arrived on 1 February, and remained till the beginning of August. The priests related that at midnight during the summer solstice the sun no more than dipped below the horizon, 'as if behind a little mound', *quasi trans parvulum tumulum*, and that the light was then sufficient for a man to do any sort of work, 'even to picking the lice out of his shirt'. Dicuil goes on to say that though it had been stated that in winter-time 'Thile'—which is the name he gives to the island—was ice-bound and shrouded in perpetual darkness, this had been definitely disproved by the experience of these travellers, who had been able to reach the coasts of 'Thile' in the 'natural period for great cold' and found there was the usual sequence of light and darkness 'save at the time of the summer solstice'. Taken in conjunction with the evidence of Scandinavian sources, it would appear from this narration that the 'Thile' of Dicuil and his compatriots was certainly Iceland.[14]

Two points about Dicuil's account are of special interest. First, there is no suggestion that this, the first voyage to Iceland of which there is certain documentary evidence, was regarded as any extraordinary achievement. Second, it is no way implied that Dicuil's informants were actually the first to discover 'Thile'. It would indeed be astonishing if a voyage of discovery had been made in mid-winter.

Though it has often been asserted that this venture of 795 was made in a curach, there is really not enough evidence to support such an assumption. Certainly, in the sixth and seventh centuries the impression one receives from considering all the available sources of information is that the curach was the usual craft for a long sea-voyage. But the great era of curach navigation had ended, in fact, well before 795. Indeed, as far back as the time of St. Columba we hear about 'long ships' made of 'hewn oak and fir' and of timber transported overseas for shipbuilding purposes.[15] Nothing is really known about the Iceland venture of 795, either as regards the type of vessel used or the course that was steered: only the bare fact that these Irishmen set out and returned in safety and that after an outward passage made in the depth of winter.

If the evidence in support of earlier Irish ventures is not entirely conclusive, there can, at least, be no doubt that the voyage of 795 was followed by later ones. Owing, possibly, to the wholesale destruction of monastic libraries during the troubled times of the Viking invasions all record of these voyages has been lost. But that communication was for long kept up with Iceland is certain. The Irish hermits were still living out there when the Norsemen began to arrive in the following century. The *papar*, as the latter called the, were for the most part settled on the south-east coast, and were stated to be 'very few in number'. There is a tantalizing allusion to other sources now, one fears, irretrievably lost. Thus: 'it is stated in English books at that time there was journeying between the lands',[16] i.e. between Iceland and the British Isles.

The navigational conditions on the part of the coast where the *papar* settlements were situated resembled to some extent those on the irregular, deeply indented shores of western Ireland. A narrow channel led between rocks and islets into the Papafjördhur. Even with quite moderate inshore winds there was frequently a heavy swell, and the strong tidal streams in the entrance to the channel would cause the sea to break heavily. In the early Middle Ages the lagoons in the vicinity were almost certainly bays and inlets. All things considered, the region in question might fairly be reckoned a curach coast, in much the same way as those parts of the Irish and Scottish littoral referred to earlier.

Some of the Irish hermits, not content with the seclusion and remoteness of 'Thile' proper, appear to have established themselves in even greater austerity on a long, low-lying, grassy island, situated about four miles off the entrance to the Berufjördhur, afterwards called Papey. There was good grazing on Papey and the surrounding islets. It is possible that the *papar* took sheep with them to Iceland, as they already had to the Faeroe Islands. Goats also might have been kept for the sake of their milk.

There was another *papar* settlement about ten miles inland at Kirkjubœr in Sida, behind the dangerous and harbourless south coast of Iceland, for long shunned by the Norsemen: in this fertile vale, backed by hills and birchwoods, it is stated in the *Landnámabók*, 'no heathen man might dwell'.[17]

What happened to the Irish hermits in Iceland when the Norsemen arrived is not at all clear. We are informed that they left the island. But when, and in what circumstances, does not appear. Did the old and the new settlers in Iceland live—at least for a short time—side by side? Were no more Irish voyages made to 'Thile'? It is implied in the *Íslendingabók* and other sources that not merely the whereabouts, but also the approximate numbers of the hermits, were known to the Norsemen. It is also possible, as Bugge has observed, that the latter, at any rate, knew enough about Irish script to recognize it at sight.[18]

Perhaps the most likely explanation is that the *papar*, whose primary object in voyaging to Iceland had been to dwell in solitude in some remote and uninhabited land, had no thought but to withdraw immediately they found that solitude broken into by the Norsemen, just as their brethren had already withdrawn from the Faeroe Islands. Between them and Naddod the Viking and his like there was a great gulf fixed.

It is apparent that the Irish eremitic activities did not entirely cease. The *Landnámabók* tells how Ásolf Aslkik, whose mother was Irish, came to Iceland with eleven companions and settled on the east side of the island. He was a Christian and held himself severely aloof from the heathen around him; from whom he would not so much as accept food. In his old age Ásolf became a hermit.[19]

We have seen how for several centuries Irish mariners had been venturing, stage by stage, out into the Atlantic. First, their navigation embraced the islands and island-groups off the western coasts of Ireland and Scotland; second, the Orkneys and Shetlands; third, the far-off Faeroes; and fourth, at some date unknown, 'Thile' or Iceland. The question which has now to be considered is, did Iceland represent the utmost limit of the Irish discoveries in the Western Ocean, or did they eventually arrive at any transatlantic region such as Greenland or lands still further to the west?

Though no positive evidence exists of such a venture, it is anyway within the bounds of possibility. In view of the long-continued sailings between Ireland and Iceland, it is by no means improbable that from time to time one of their craft would be forced off its course by a gale, and so driven westward to Greenland. In later centuries this was to happen to quite a number of Iceland-bound merchant-men,[20] and it may well have happened to some of the lighter and less weatherly craft used by the monks.

There is a certain passage in Dicuil's account of the voyage of 795 which is highly suggestive. Dicuil records that these mariners had, a day's sailing north of 'Thile', discovered frozen sea.[21] It is evident from this that on at least one occasion an Irish craft had sailed on round the Iceland coast, and ventured for some distance into the open sea to the north of the island, where they had apparently encountered drift-ice. Since most of the *papar* settlements were situated on the south-east coast of Iceland (where the astronomical conditions described in Dicuil's narration are, in fact, precisely the same as those obtaining on that part of the coast), it is clear that the shortest route thence to the north of the island would be 'east about'. Against this, however, it is to be remembered that the current, which sets in a clockwise direction round the coast of Iceland, would be contrary. It is therefore possible and even likely that these Irish in 795 circumnavigated the island; and if on this occasion, then probably on others besides; and, if it so happend that they drifted far enough out into the sea to the westward, they may conceivably have sighted the distant mountain peaks of Greenland. As has been suggested in connection with the discovery of Iceland, there is also the possibility of a mirage leading them on to explore further.

There are also various significant passages in the legendary voyages of St. Brendan and others which, as has already been explained, suggest that in very early times Irish seafarers may have had some knowledge of the far western lands which were later to be discovered or rediscovered by the Norsemen. In the same way that the 'Isle of Sheep' has been identified with the Faeroe Islands, and the volcanic isle with Iceland, it has been claimed that St. Brendan's 'Pillar of Crystal' was probably an iceberg (only to be encountered on the other side of the Atlantic), that his 'black fog' and 'great shoal of fish' suggest the banks of Newfoundland, his 'swarthy dwarfs' the Eskimos of Greenland, and the fearsome monster which had 'tusks like a boar' the walrus. The difficulty, however, is to know exactly at what date the different elements of the Brendan legend originated; and in the absence of such knowledge no reliance is to be placed on this kind of testimony.

How far in fact the Irish may have roamed in the wastes of the North Atlantic must remain a matter of conjecture. Certainly they discovered all the isles to the north and west of these islands. Did they ever sight Rockall? It was well within the range of the old sailing-curach. Sooner or later they must have discovered that the ocean to the southward of the Faeroes and Iceland was empty of islands. Where else may they have sailed? It has been suggested that the Porcupine Bank, situated some 140 miles west of Ireland, was known in very early times to deep-sea fishermen. In 1934 a fragment of a Roman *olla* was dredged up in the trawl of the *Muroto*.[22] Too much, however, must not be argued from this. All that it proves positively is that at some unknown period a craft carrying Roman pottery must either have sunk, or passed, over the Porcupine Bank. Beyond that all is conjecture; though probably the most likely explanation of the presence of Roman pottery on the Porcupine Bank is that some war-band from the west, returning home in their curach with their booty after a raid, were blown far out into the Atlantic and wrecked there.

The Irish navigation to the Faeroes and Iceland was clearly no 'hit or miss' affair. The Irish settlements in the Faeroes lasted from about 700 to 800. Those on the south-east coast of Iceland lasted for at least eighty years; and probably for even longer. Whoever was responsible for the navigation to Iceland—whether he were monk or mariner—evidently knew his work. The long continuance of these voyages gives testimony of a standard of seamanship and navigation comparable, one may perhaps believe, with that of the Norsemen of the early Viking era.

In none of the Irish sources, admittedly, are there any passages resembling the seamanlike narrative of the voyage of Bjarni Herjólfsson in the *Grænlendinga saga*. Compared with the well-documented voyages of the Viking age the ventures of these monks and hermits are singularly devoid of professional detail and interest. The *immrama*, indeed, are far more concerned with marvels and miracles and meditations on the Four Last Things than with the prosaic business of deep-sea navigation.

Nevertheless, though there is little enough to be gleaned concerning seamanship and navigation in the early Irish sources, what little there is is not unimportant. First, it would appear that in the larger craft, such as the sailing-curach, it was customary for the clerics to be accompanied by professional seamen, who are styled *nautae*; mention is also made of a pilot, *gubernata*.[23] Considerable numbers of experienced seamen must have been needed to man the large war-fleets which from time to time were launched against the Picts and others. Similarly, in the sixth century and later a good many mariners were required for the numerous voyages made by the monks of Iona and other communities. Whether a cleric or a seaman acted as master is unknown.

The Irish seem to have had little fear of the open sea. This is implied in a number of passages; on at least one occasion mention is made of their proceeding direct to Tiree from Iona instead of hugging the land.[24] At a later date, the Irish also appear to have sailed direct to Iceland: it is neither stated nor implied by Dicuil that his informants sailed to Iceland by way of the Faeroes. The reference to the 'direct course' shaped by Cormac ua Liatháin and his crew on his third voyage in search of the 'desert in the ocean' is of particular significance; for in view of the fact that their curach was far from sight of land at the time, it is evident that they must have steered with the help of the heavenly bodies.[25] There was formerly a tradition in the Faeroe Islands that the Irish monks used to make the Iceland voyage early in the year so as to avoid the fogs and 'luminous nights' of later months.[26] (From about the middle of May to the beginning of August in the high latitudes the stars are invisible.) All this serves to support the theory that the Irish depended upon the sun and other stars for guidance across the immensity of the ocean.

We also learn how the mariner became aware of the proximity of land by listening to the sound of the sea on a strand, *tuindi fria tracht*, and by noting the flight of birds, like the falcon flying to the south-east in the *Immram curaig Maíle Dúin*.[27] (It is worth noticing that passages similar to these are to be found in the Icelandic sagas.) Of necessity they must often have had recourse to all kinds of adventitious aids to navigation, which were particularly important when they were approaching a dangerous coast in thick weather or fog.

The meagre details concerning seamanship supplied by the early Irish sources may be supplemented to some extent by the experience of the Irish curachmen of modern times, and also by that of the Norse seamen of the Viking age who for their ocean passages used a light and handy merchant vessel resembling, in certain respects, the Irish sailing-curach. Even more than the Norse merchantman, the sailing-curach had the defects of its qualities. It was liable to make much leeway under certain conditions of wind and sea, and, like the Norse merchantman, must have been driven for scores, and possibly even hundreds, of miles in heavy gales. As in the Viking vessels, the principal duties of the crew must have been managing the large sail, steering, baling, and standing look-out. In severe weather everything depended upon the skill of the helmsman in utilizing the buoyancy and manœuvrability of the curach to avoid a dangerous sea. Then as always, this kind of skill was the product of long years of seafaring experience. In modern times a crew would sometimes in an emergency run their craft ashore on an exposed strand in the hope of saving life, even if they lost the curach : choosing, if they could, the best spot for running in, and listening at night for the interval between the fall of two seas to gauge their strength. It is more than likely that in desperate straits their ancestors would resort to the same manœuvre, which was also, as we shall presently see, familiar to the Norsemen and was styled by them *sigla til brots*, 'running ashore under sail'.[28] Like the Norsemen, they would be well aware of the danger of roosts and tide-rips,[29] not only off the Irish coast, but also in the Orkneys and Shetlands, in the Faeroes, and in Iceland. Like the Norsemen, they must occasionally have lost their reckoning and all sense of direction, as did the Aran Islanders in recent times, caught 'out at the back' in fog or thick weather, only to find themselves presently rowing in a circle. Last but not least, long experience of the weather and signs of the weather in the seas they were traversing would have been all important to the crew of a small sailing-vessel.

The question naturally arises, did the Irish transmit any of their knowledge of these matters to their Norse successors?

Of positive proof there is virtually no trace. Little is known, indeed, of the earliest Norse voyages to the Faeroes and Iceland. We cannot say in what circumstances the first shipload of Norsemen reached the Faeroes: and the earliest and most trustworthy authority on the settlement of Iceland—the *Íslendingabók* of Ari Frodi—is silent as to *how* Iceland came to be discovered by his compatriots. However, this much may safely be said. There is, at least, an equal possibility that the two sea-routes, both of them familiar to Irish mariners for several generations, were by them made known to the Norsemen, with whom they had been for some time in close contact. It is always to be remembered that some of the first Vikings to come to Iceland seem to have followed in the track of the Irish pioneers. Thus Gardar Svávarsson, who had lived in the Celtic Hebrides, arrived in that region of the Iceland coast—the south-east—where the *papar* were established; and Naddod, 'who was a great Viking', sailed from Norway to the Faeroe Islands and thence to Iceland—apparently he knew not only the nearest land to Iceland, but also the approximate direction in which Iceland lay. It would be rash, in fact, to assume that the knowledge and experience which rendered possible the ocean voyages of the

Viking era were of purely Scandinavian origin; rather is one inclined to doubt whether the interesting instance of the Irishmen's sea-lore being passed on to their Norse shipmates, as related in the *Landnámabók*, is altogether an isolated case; and one would like to know more about the Christian man from the Sudreyjar who composed the verses about the dreaded *hafgerðingar*.[30]

To sum up: in the course of centuries the Gaels had evolved a type of sailing-vessel—handy, buoyant, and eminently seaworthy—which was capable of regular passages up and down the west coast of Ireland and Scotland, and to outlying islands such as the Arans and Iona, and even, on occasion, of voyages far out into the ocean. They had acquired a large and varied stock of sea-experience which eventually culminated in a series of Irish ventures to Iceland. Though their method of navigation is unknown to us it is certain that they were able to voyage for several hundred miles in the open sea without the aid of the magnetic compass, which actually did not come into use, in the North, for another five hundred years. The Iceland ventures continued until well on in the ninth century, when, on the arrival of the Norse settlers, the Irishmen apparently quitted the island. These voyages constitute the first true ocean passages ever achieved in northern waters, extending nearly half-way across the Atlantic.

II

THE OCEAN TRAFFIC
OF THE NORTH

6

The genesis of Norse expansion

ACCORDING TO THE chroniclers the Viking raids fell with startling and dramatic suddenness on the coasts of Europe. They are first heard of in the closing years of the eighth century. It has been said that the terror and abhorrence they evoked caused a new petition to be added to the litany—'ab ira Normannorum libera nos, Domine'.

Many factors contributed towards this sudden and astonishing outburst of energy which manifested itself in the Viking age.[1] The growth of a surplus population in many districts of Scandinavia;[2] the aggression of Charlemagne in northern Germany, and the consequent threat to Denmark; the reduction of Frisia—hitherto the leading maritime and commercial power in the North—by the Franks; the ambition of individual chieftains and their sons; the steady progress of shipbuilding and seafaring activities. To these may be added: the greed for plunder, the love of adventure, and, perhaps strongest of all, some unknown *motio valida*, some secret and powerful urge[3] like that which inspired the Crusades. The Scandinavians were now the leading maritime people of the Continent. The coasts and ports of western Europe lay open and defenceless. Two or three days' sail across the sea were lands, loot, and slaves: theirs for the taking.

The various political and economic factors need not be considered in detail; but the development of shipbuilding, seamanship, and navigation is of crucial importance at this stage, since, if this had not occurred, much of the overseas expansion of the Viking era would have been impossible.

The progress of shipbuilding in the North is closely associated with the development of the Norwegian iron industry. This made possible the production of an abundance of iron tools, of which the iron axe was the most important. With these tools, and with the almost illimitable supplies of timber available from the forests of Norway, the Norsemen attained a remarkable proficiency in house- and boat-building. The sea-going sailing ship of the Viking age is the outcome of a long and gradual process of development, the successive stages of which can be reasonably well established by means of the Valderøy, Hjörtspring, Nydam, Kvalsund, Oseberg, Gokstad, Tune, Hedeby, Ladby, and other vessels which from time to time have been disinterred in different parts of Scandinavia.

According to Shetelig, the prototype of the Scandinavian plank-boat is the Danish Hjörtspring craft. Its planks are sewn together with a cord, and the stitches and seams caulked with resin; the scantlings are slender hazel branches fastened to the planking by lashings through cleats contrived in the planks. It was propelled by paddles. The Hjörtspring boat is of comparatively weak construction, and could scarcely have been a sea-going craft. There is no trace of any fitting for mast and sail.

It dates back to about the third century B.C. The Valderøy vessel is believed to be still older.

The Nydam craft, which dates from the early fourth century A.D., is clincher-built with iron nails and double-ended, with considerable sheer and high stem and stern-post. It has an over-all length of about 76 feet; its beam is $10\frac{1}{2}$ feet, and its depth about 4 feet. Incorporated in this vessel is the distinctive principle of construction which may be observed in the Hjörtspring boat some six centuries earlier, and which is also applied to the sea-going craft of a later era. The planking which forms the vessel's skin is lashed, not riveted, to the frame timbers; the ribs rest on lugs or cleats which project from the planks; and the cleats are bound to the ribs with bast cords. The Nydam craft is not fitted with a true keel, but with a broad plank projecting only very slightly below the bottom of the vessel. A steering-paddle is lashed to the starboard quarter. There is no sign of a mast. The Nydam craft was propelled by fifteen oars on each side. It was a rowing-boat pure and simple.

The Kvalsund craft, which dates from about the beginning of the eighth century, may be said to represent the half-way stage between the Nydam boat and the fully developed longship of the Viking age. It is nearly 60 feet in length, with a beam amidships of about 10 feet and a depth of rather less than 3 feet. There is, however, an important advance in respect to both the hull and the side-rudder. 'The keel plank with an external reinforced fillet comes close to being a true keel. The hull is more stable than that of the Nydam Ship, being fuller amidships. This ship could have been sailed, but as no traces of any attachment for a mast remain, it is impossible to say whether it was actually used for sailing.'[4] It marks, therefore, an important point in the development in the North of the sea-going sailing-ship. The Kvalsund craft has also far more substantial frame timbers; moreover, its beam is greater in proportion to its length, thus giving the ship greater stability. The Kvalsund craft has 8 strakes as compared with the Nydam boat's 5; and each strake consists of several pieces. Moreover, whereas the lower planks are secured to the ribs with lashings, the planks above the water-line are fastened with wooden nails. Further, the rudder is stepped to the starboard quarter by a device which, while giving the blade the requisite play, keeps it clear of the vessel's side. It is similar to the rudder of the Oseberg vessel.

The final phase of these developments was reached somewhere between the time of the Kvalsund and that of the Gokstad vessel: that is, at some time during the late eighth or early ninth century. It is apparent from the Gotland sculptured stones that vessels which in some respects resembled the Gokstad craft—being fitted with mast and square-sail with reef-lines and various appurtenances—were already at sea in the eighth century.

The Gokstad craft is approximately 80 feet in length, with a breadth of nearly 17 feet and a depth amidships of nearly 7 feet. It is built entirely of oak, with a deep external keel and high stem and stern-post. The planking consists of 16 strakes on each side, riveted together clincher-fashion, the 8 lower strakes being lashed to the ship's frames, or ribs, by withies. Such a method of fastening, which allowed of a very much thinner planking than would have been possible with the strakes simply nailed to the ribs, was in large measure responsible for the extraordinary buoyancy and elasticity of the Viking craft: the whole structure could give without breaking.[5] The frames themselves are not fastened directly to the keel (but only indirectly by

Memorial stone with carving of a longship from Larbro̊, Gotland

means of the garboard strakes) and rest loosely in shallow grooves on the keelson. The vessel was propelled by both sail and oars. The single mast, which was stepped amidships, was probably lowered when the oars were in use. The ship pulled 16 oars a side, each some 18 feet in length; these oars were worked through small circular oar-ports cut in the vessel's side some 18 inches below the gunnel.

The side-rudder, secured to a wooden boss on the vessel's starboard quarter by means of a withy, is controlled by an athwartship tiller, *hjálm*. To allow the rudder to turn on its vertical axis the axle had to possess some degree of elasticity. (In the Oseberg vessel the axle, *stjórnvið*, was made out of a strong piece of fir-root.) It is to be noted that with the side-rudder both sides of the rudder are effectively in action. It also acts to some extent as a keel.[6]

Nailed to the vessel's sides abreast of the mast are two blocks of pinewood. It was for long a matter for conjecture what purpose these blocks served. It is now established that they served as a support for a spar called the *beitiáss*, or sprit-pole. One end of the *beitiáss* was inserted in a socket in one of these wooden blocks and the other was placed in a cringle in the weather leech of the square-sail in order to prevent the sail from shaking while sailing on a wind; thus enabling the vessel to sail closer to the wind. Without the *beitiáss* the ship might sail with a beam wind, but would not be able to 'bite' into the wind—i.e. beat to windward. The maximum close-hauling which could be achieved by this means was called *at aka segli at endilöngu skipi*. The *beitiáss* was also used to spread the sail when sailing with a following wind. The disposition of the sockets in the wooden blocks shows clearly that the *beitiáss* was thus used when running free, as well as when sailing on a wind.[7]

Unlike its forerunners, the Gokstad vessel was fitted with a proper keel. It now became possible for these northern craft to face gales and heavy seas: to sail with a beam wind, or even nearer to the wind. The Gokstad ship is beyond comparison the finest and best preserved of the Viking craft which have come down to us through the ages.

The Tune ship is a large, open, clincher-built craft closely resembling the Gokstad vessel in design and construction. It is about 67 feet in length, with a beam of 14 feet, and 4 feet depth. Its great breadth in proportion to its draught may perhaps be accounted for by its having been adapted to special local conditions. The frame timbers were lashed to cleats contrived in the planks with bast cords. The stump of a strong mast still survives. The Tune craft was in all respects a proper sea-going ship like the Gokstad craft.

The remarkable advance in ship-design ushered in the new age of Scandinavian maritime enterprise and expansion. The Gokstad and Tune vessels date back to about the mid-ninth century—the heyday of Viking raid and foray.

The Viking ships were for the most part built of oak; but according to the sagas birch, beech, and fir were also used. The Gokstad and Oseberg ships were built throughout of oak. The Tune ship was built of both oak and fir. The keel and stem were always of oak. Some of the finest oak groves in Norway today are situated in the Vestfold, where the Gokstad and Oseberg ships were discovered. Such a trunk as was required for the keel of the Gokstad ship, however, could scarcely be found in Norway at the present time. In 1893 the trunk for the keel of the *Viking* (a replica of the Gokstad ship) had to be imported from America. There would probably be no lack of suitable oak in the coastal regions of Sogn, Hordaland, and Rogaland,

and particularly in such thickly wooded districts as around Ballangen, Lavangen, Salangen, and Varanger fjords. The heavy drain of timber fit for shipbuilding continued throughout the ensuing centuries.[8]

It was the Norse mastery of ship-design and construction which made possible the first Viking raids towards the close of the eighth century. It was in the course of that century, probably by the middle decades, that the true sailing ship had been evolved. The significance of these developments in the history of the North can scarcely be set too high. 'For four hundred years the ship is the vital thread in the history of Norwegian unification. It is much more a history of naval warfare than of politics and diplomacy. Everything from the battle of Hafrsfjörd to the battle of Largs (1263)—the last of the old Norse empire—has its share and background in the ship.'[9]

Behind the plundering raids and ocean ventures of the Viking age lay uncounted centuries of seafaring experience. We know from the evidence of archaeology that even in prehistoric times the peoples of Scandinavia had always turned to the sea, which was by far their most important, and—in a great many parts of the country—almost their only, highway. In this matter geography was the governing factor. It has been calculated that when the fjords and other inlets have been taken into account Norway possesses a coastline of well over 20,000 miles. It follows from this that the mass of her population has always lived within easy reach of the sea. While the fjords and mountain barriers made travelling by land always difficult and often impossible, a busy traffic in course of time arose, up and down the western coast of Norway, in the sheltered channels (*skergarðr*) behind the long chain of islands and skerries. 'This fairway', observes Brøgger, 'is the main street of Norway from the south to the north, a highway as significant in the history of our country as the great rivers in that of Europe. It is the chain which links the different parts of Norway together, and is undoubtedly the most prominent feature of the history and civilization of the country.'[10] Similarly, the Kattegat, with the Sound, was a link, rather than a barrier, between the contiguous parts of Norway, Denmark, and Sweden. The navigational conditions obtaining on both the western coast of Norway and the Baltic coasts of Denmark favoured the use, in very early times, of comparatively weakly constructed craft which could scarcely have ventured across the open sea.

The Atlantic fishery was one of the major factors in the early development of Norse seamanship and navigation. In the Hordaland-Sogn region, and also in the high North, by the Lofoten–Vesteralen island-groups, the banks lay in convenient proximity to the land. From time immemorial fishermen had been accustomed to rely upon landmarks to guide them to the fishing-grounds. The same landmarks and bearings had been used by the coastwise shipping on its way up and down the *skergarðr*. Fishermen must also have learned a great deal about the local tidal streams and currents; the bearings of rocks, reefs, and shoals; the approximate position of deeps and banks; and the habits of local fauna (especially seabirds). In the course of centuries their knowledge of the geography of the North Sea—of the deeps like the Norwegian Channel, the Long Forties, and the Gut; of the relatively shallow areas like the Viking Bank, Bergen Bank, Dogger Bank, and Jutland Bank—must have little by little increased. On their threshold was the Norwegian Channel leading down the west coast of Norway from the latitude of Cape Stad to the entrance of

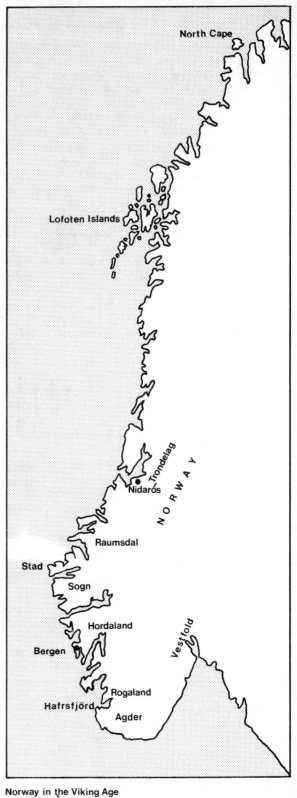

Norway in the Viking Age

the Skagerrak. Experienced mariners would often be able to tell by the colour of the water and run of the seas when they were in this channel and when they had entered the plateau of the North Sea. Similar signs would also show when they were over one of the banks. The run of the seas would often give the mariner some rough idea of direction. With westerly to southerly winds the long westerly seas are to be expected. With northerly to easterly winds the seas are as a rule steeper and shorter. In certain conditions of wind and cloud it is sometimes possible at dusk to distinguish east and west by the more luminous appearance of the latter. Again, the flight-line of certain seabirds would often be of assistance when the ship was in the vicinity of cliffs or rocks where the birds were breeding. The rich fishery off the western coast served as a school of hardy seamen. We know from *Orosius* that whale and walrus fisheries were carried on in the Nordland as early as Ohthere's time.

It is important to remember, too, that throughout the whole of the Viking age trading expeditions were sent out as well as marauding war-fleets.[11] Thus the Swedish Vikings were not only warriors, they were also enterprising traders. They trafficked as far south as the Black Sea and Constantinople. From the early Middle Ages Gotland was an entrepôt of Baltic trade. It was the meeting-place and mart of two civilizations: Flanders and western Europe on the one hand, and the immense Russian plain on the other. During the Viking age frequent mention is made of merchant-mariners; such as Hrafn the Limerick trader; Högni, who inherited the ship *Stígandi* from his father, Ingimund; Thórir, who had a *knörr* built in the forests of Sogn; Björn, the brother of Eirík Blood-axe; Hrafn the Novgorod trader, and many others.[12]

Norsemen were crossing the North Sea at a considerably earlier date than that of the first archaeological evidence of their settlements in the Orkneys and Shetlands. Raiding and trading expeditions to the countries 'west over the sea', *vestan um haf*, apparently preceded, as one would expect, the first voyages of emigration: in the same way Viking crews had visited the distant Faeroe Islands some time before the first permanent settlements were made there. We know that they had reached the Faeroes by about the end of the eighth century. Such a passage would scarcely have been possible without a good deal of previous experience of voyaging to and fro across the North Sea. It is perhaps premissible to conjecture that Norsemen were occasionally trafficking with the lands *vestan um haf* by about the year 700.[13]

The passage to the Shetland Islands from the fjords of Hordaland would not have been a very formidable undertaking for Norse mariners of the eighth century. From Bergen to Yell is about 230 miles. In spring, when easterly winds are to be looked for in the North Sea, the crossing would not take more than two days and two nights. The men would keep careful watch on the coastal mountains, as Norwegian fishermen do this day, to bring out the familiar *mið*—the fix which would give them their points of departure. They must steer due west as nearly as they could, so that they would have the Pole-star on the starboard beam during the hours of darkness, and the sun on the port beam at midday. Some fifteen hours after sailing they would be over the Viking Bank, when a significant change in the colour and run of the sea would often give them some idea of their position. Finally, as they approached the out-lying skerries they must keep a sharp look-out for small auks and other local seabirds which might give them a rough check on their course.[14]

In the ninth century, and perhaps even earlier, successive waves of emigrants sailed away to the Shetlands, Orkneys, and Caithness. The peasant farmers of Agder and Rogaland appear to have formed the backbone of this emigration. Brøgger has argued that the colonization of the Orkneys and Shetlands began 'long before the year 800'. But adequate evidence of this is lacking. At any rate by about 800 the Norsemen had certainly begun to settle in the islands to the north and west of Scotland. The settlers encountered no effective resistance, and their occupation of the islands was apparently accomplished without any violent struggle.[15]

The first that is heard of Viking raids to the westward is on the occasion of the sack of Lindisfarne, in 793, by some Vikings from Hordaland. After that the exodus proceeds with ever-increasing momentum. It was this *vestrvíking* which was the main and governing factor in the development of Norse shipbuilding, seamanship, and navigation. In 794 the raiders sacked the monastery of Wearmouth. In 795 they fell on Lambay near Dublin; and also on the coast of Wales. In 799 they were in the Bay of Biscay. In 802 and again in 806 they sacked Iona, and in 807 Inishmurray. The raiders were as yet few in number—perhaps only two or three ships at a time; and their operations were on a relatively limited scale. A few years later, however, far stronger fleets appeared, and large and well-armed bands carried sword and fire well into the interior. 'Out of the north,' wrote Alcuin, quoting appositely from Jeremiah, 'out of the north an evil shall break forth upon all the inhabitants of the land.'

Between 820 and 830 they ravaged Ireland from end to end. 'The Northmen', declared the *Book of Chronicles*, 'left no saint or scholar, or noble church, or [hermit's] cave, or island unplundered in Ireland.'[16] The Viking fleets sailed or rowed up the great rivers and anchored in the inland loughs; the larger churches and monasteries with their accumulated treasures were their principal target; the Vikings pillaged every great monastery in the island—a few of them several times over. The harassed monks were thankful for the winter storms which granted them a short respite from these merciless attacks.

> Bitter is the wind tonight,
> It tosses the ocean's white hair:
> I fear not the coming over the Irish Sea
> Of fierce warriors from Norway.[17]

From their Irish bases the Vikings appear to have attacked various places on the Atlantic coast of France. In 844 a large Viking fleet sailed southward around Cape Finisterre to Lisbon, plundering as it went. In 845 Ragnar Lodbrók sailed with 120 ships up the Seine and sacked Paris. In 851 350 Viking ships arrived in the estuary of the Thames. In the next generation or so they occupied the Danelaw in England and the French province that we now know as Normandy. While the Vikings of Norway and Denmark were ravaging in the west, the Swedes were active in the Baltic and in Russia. They occupied Novgorod and Kiev about 865, and later sailed down the Dnieper and Volga to Constantinople and Astrakhan. Swedish war-fleets cruised and pillaged off the coasts of the Black Sea and Caspian. The Swedish penetration of Russia reached its peak in the course of the tenth and eleventh centuries.

By this time the Scandinavians, ranging ever further afield, had fought, burnt, looted, and conquered on almost every coast in Europe. Their longships had appeared on the Elbe, Rhine, Humber, Thames, Severn, Shannon, Seine, Loire,

Garonne, and Guadalquivir. Atlantic trade and settlement had gone hand in hand with Viking foray and pillage. They had colonized the Sudreyjar, or Hebrides, as well as the Orkneys and Shetlands, various parts of Ireland, and much of the Scottish mainland.

In the ninth century the Norsemen had set out across the ocean which was described in the later words of Adam of Bremen as *terribilis et periculosa*. The Viking age in fact constitutes the first great era of overseas expansion in the history of the West. The ocean navigation of the Norsemen may be said to have begun, early in the Viking era, with the settlement of the Faeroe Islands. Before the turn of the century the colonization of Iceland was fairly under way. The last great surge of Viking enterprise and expansion came at the end of the following century, resulting in the discovery of Greenland and parts of North America. After that the movement gradually subsided.

It is surely significant that, according to the *Landnámabók*, so many of the chieftains and others who ventured to Iceland at the time of the settlement were of Viking stock. Moreover, the majority of them were from Norway. Neither the Danes, looting in Frisia and Ireland and carving out for themselves principalities in England and France, nor the Swedes, cruising down the great Russian rivers to barter the produce of the North for the luxuries of the Orient, ever earned particular renown as deep-watermen. The long cruises made by the Vikings of Hordaland and the neighbouring provinces into the stormy Atlantic must certainly have acted as a forcing-house both of seamanship and warlike skill. The range and scale of these operations stamped the Norsemen as practised masters of navigation and of war.

7

The emigration to the Faeroe Islands

IT IS RELATED by Snorri Sturluson in the *Heimskringla* that the Faeroe Islands were first discovered and settled in the reign of Harald Hárfagri.[1] This, however, is an obvious anachronism. The Norsemen, as is apparent from the testimony of the Irish geographer, Dicuil, were already sailing to the islands by about the year 800. Their appearance was followed by the departure of the Irish clerics who had been settled on the Faeroe Islands since the beginning of the eighth century. It was these earlier settlers who had stocked the islands with the sheep which still abounded there on the arrival of the Norsemen.[2] A runic inscription, discovered at Kirkjubœr, has been ascribed to about the year 800. According to the *Landnámabók*, Naddod, a Viking from Agder, who is mentioned as one of the early voyagers to Iceland, was already living on the Faeroe Islands, where he must have settled about the middle of the ninth century. Also according to the *Landnámabók*, the grandson of Grím Kamban (who is stated to have been the first Norseman to settle in the islands) arrived in the Faeroes in the early years of the same century.[3] There can be no doubt but that the first voyages of the Norsemen to the Faeroe Islands preceded by more than two generations their discovery of Iceland in Harald Hárfagri's time.

As was to happen later in Iceland and Greenland, the Norse expansion to the Faeroe Islands appears to have been accomplished in three fairly distinct stages. First, there was the discovery of the islands by the Norsemen. Second, there was a series of voyages in the course of which the explorers got to know the lie of the land. Third, there was the actual settlement. The emigration to the Faeroe Islands marked a further stage in the westward expansion of the Norsemen, and came in fact as a sequel to the *vestrvíking*.[4]

It is apparent that one of the major causes of the emigration to the Faeroe Islands was the steadily increasing population of Norway at the time. The archaeological evidence shows fairly clearly that in the course of the eighth century many of the farms situated along the west coast of Norway became over-populated.[5] The surplus either moved into the interior or went overseas. Some of these emigrants made their homes in the fertile lands of eastern England and Normandy; but by far the greater number sailed westward across the sea to new settlements in the North.

The choice of northerly regions by the majority was due to fundamental and primitive causes. It was the craving for surroundings where something of the old was to be found in their new activities. They asked for sea and fjord, mountain and hill, the fowling cliffs and sealing grounds. They needed the pastures, meadows, and heather, to which they had been accustomed in the land of their birth, and the light summer nights which brooded softly over farm and field in Norway. No sentimental spirit of homesickness lay at the back of all this, but the simple fact that the whole of their mentality, fostered by the toil of countless generations before them, was adjusted to a life in which all these things were to be found. All else would be in the nature of transplanting, obliteration, and sacrifice. It would deprive them of the powers which were their inheritance and their greatest asset.[6]

From a careful consideration of the place-names, runic inscriptions, and other archaeological evidence, it is possible to form some idea of the origins of a good many of these settlers. It is evident that there were two main streams of emigration: one of these streams proceeded from the coastal district between East Agder and Hordaland, and the other from the islands *vestan um haf*, 'west over the sea'. The earliest of the runic inscriptions which have come to light in the Faeroes—namely, the stone at Kirkjubœr—closely resembles contemporary inscriptions in Rogaland.[7] The emigration was carried on under the leadership of chieftains on much the same lines as the later settlement of Iceland. Whether there were any Celtic inhabitants living in the Faeroes when the Norsemen took possession of these islands—and, if there were, what was their fate—remains obscure.[8] It is probable that from the earliest years of the emigration (as was unquestionably the case at a later period) the sheep occupied a dominating position in the economy of the islands. Wool and wadmal were staple exports of the Faeroes throughout the medieval era.[9]

From the standpoint of maritime history the beginning of this traffic to the Faeroe Islands is of crucial significance. The passage of all these emigrant families with their farm and household gear and personal possessions across three hundred miles and more of the open Atlantic was alike a portent and a turning-point. The Norse emigration to the Faeroe Islands formed a prelude to the far-flung oceanic expansion of the Viking age; and marked an epoch in the evolution of the shipping, seamanship, and navigation of the North.

Six centuries before the English, Germans, or Dutch, the Norsemen had evolved a craft which was admirably adapted to the navigation of the Western Ocean. Hitherto none of the vessels they built were capable of carrying what were substantial cargoes on such passages. All things considered, it is probable that some kind of ocean-going merchantman, or *hafskip*, was already in use before the year 800.

The vessel used for these earliest ventures to the Faeroe Islands was presumably a primitive type of *knörr*,[10] embodying many of the constructional features of the Gokstad craft.[11] The *knörr* is mentioned in the oldest Scandinavian poetry as well as in the *Landnámabók*. Two *knerrir* are recorded to have fought at Hafrsfjörd.[12] The voyages to the Faeroes, together with the trade to the west coast of Ireland, must have played an important part in the further development of the *knörr*. Just as the traffic with the Scottish Isles had prepared the way for these ventures to the Faeroes, so would the latter in turn prepare the way for the sailings to Iceland during the later part of the ninth century. The opening up of the new sea-routes reflected the steady progress of Scandinavian shipbuilding.

No less did the long journey to the Faeroe Islands mark a significant advance in the navigational knowledge and experience of the Norsemen. They had to memorize new bearings, courses, and distances; they had to contend with another set of surface currents; they had to become familiar with a wider range of adventitious aids to navigation. They had to master the new technique of sailing for days across the open sea, and of spending several nights afloat out of sight of land.

The necessity of knowing the signs of the weather on this long crossing was naturally more urgent; also there was greater need for them to keep a good reckoning and to hold to the proper course—otherwise they might miss the Faeroes altogether. They had to seek out the safe anchorage and landing-places in the new colony and to discover the dangers in the approaches. Owing to the frequency of fogs in the vicinity of the islands, their growing familiarity with the adventitious aids in those waters would be especially important.[13] There are passages in the *Landnámabók* and elsewhere which suggest that, in the ninth and tenth centuries, these lessons were being learned.

It was essentially a matter of practical experience. In the early days of deep-sea navigation the chief necessity for crews making an ocean passage was, after all, *to have done it before*. The mariners of the Viking age depended upon a species of skill which has today almost vanished from the sea. In their case, inborn knowledge and experience largely filled the place of instruments and mathematics.[14]

8

The settlement of Iceland

A GOOD DEAL of obscurity surrounds the first arrival of the Norsemen off the coasts of Iceland. The honour of discovering—or, more truly, of rediscovering—that distant land has been ascribed variously to a Swede called Gardar Svávarsson, to Naddod the Viking, and to certain merchants.[1] According to the earliest and by far the most reliable authority for this period, Ari Fródi, his compatriots began to colonize Iceland at some time during the 870s and the first of them to settle there was Ingólf. 'At that time Iceland was covered with forests between mountains and shore', wrote Ari. 'Then Christian men whom the Northmen called Popes (*papar*) were here; but afterwards they went away, because they did not wish to live here together with heathen men, and they left behind Irish books, bells, and crooks. From this could be seen that they were Irish.'[2]

The *Landnámabók*, which was in part the work of Ari Fródi, though somewhat inaccurate as to certain details, nevertheless gives a sufficiently reliable report of the general outline of the Settlement. Ingólf's expedition to Iceland was followed by a great many others of the same kind. By the final decade of the ninth century the thin trickle of early emigrants (which included Ketill-Hæng, Kveld-Úlf, Skallagrím, and Grím the Hálogalander) had swollen to a broad and steady stream. The *Landnámabók* states that many of the chieftains came to Iceland in their own ships. From this and other evidence it would appear, as one would naturally expect, that a large number of the settlers belonged to seafaring families. Practically all of them hailed from coastal districts in Norway and the lands *vestan um haf*, 'west over the sea', i.e. Ireland and the Scottish isles.

Many of the freedmen and slaves were Irish. Some of the chieftains were of mixed Norse and Irish stock. Such were Vibald and Askell Knokkan, Baug, Thorgrím Grímólfsson, Helgi the Lean, Oláf 'Peacock', who piloted Örm's ship to Iceland, and Snorri Thórdsson, grandfather of Thorfinn Karlsefni. In the earlier stages of the Settlement especially, Irish and Norse tradition and experience were closely intertwined. The ancient Irish custom of building a lodge, or *bruiden*, at the cross-ways, where refreshment was freely dispensed to wayfarers, was reproduced in Iceland. Mention must also be made of the mysterious story of Hvítramannaland, or 'White Man's Land', which was a considerable Irish settlement believed to lie westward across the ocean 'six days' sailing from Ireland', not far from Vinland the Good (of which much more will be said later). From time to time various theories have been advanced to explain how the story arose; none of them particularly convincing.[3]

A large proportion of the settlers came from the west coast of Norway: from Lofoten, Hálogaland, and Trondelag down to Rogaland and Agder. A smaller

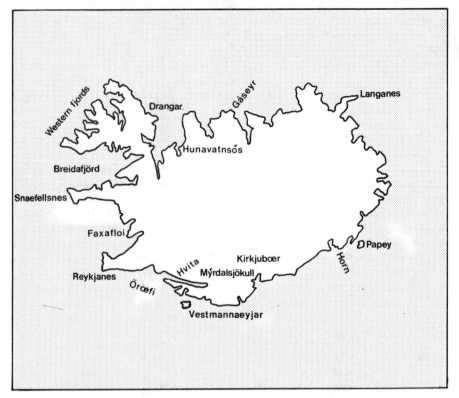

Iceland in the Viking Age

number—possibly thirty in all—were Swedes; and there may have been a few Danes from the Orkneys. Like the colonization of the Faeroe Islands the emigration to Iceland was, above all, aristocratic in character. Several of the settlers were the connections of kings and earls. Thórd Vikingsson is said to have been a son of Harald Hárfagri. Ævar was the grandson of Harald Goldbeard, King of Sogn. Sléttu-Björn was descended from the kings of Uppsala and Novgorod. One of the earliest and wealthiest of the settlers was Queen Aud, widow of the Norse king of Dublin. Ketill-Björn from Naumdale was the grandson of Earl Hákon. Hrollaug, who went to Iceland on the advice of Harald Hárfagri, was the son of Rögnvald, Earl of More. Brúni the White was the son of Earl Harek. Ketill-Hæng was also an earl's son. Asbjörn was the son of a chieftain of Sogn. Ketill-Thrym was a renowned chieftain of Raumsdal. The first chieftains to settle in Iceland held very large estates. Occasionally they gave away or bartered part of these lands to later emigrants.[4]

It has already been said that many of these emigrants to Iceland came of seafaring stock;[5] and to this it must be added that many of them were Vikings, or near akin to Vikings. From the outset the Iceland venture had been intimately connected with Viking enterprise and expansion. Thus in the *Landnámabók* reference is made to

Naddod the Viking and to Raven-Flóki, 'who was a great Viking', two of the early discoverers of the country, and to Hjörleif, who went on *vestrvíking* before he accompanied his brother-in-law, Ingólf, to Iceland; and to Ölvir barnakarl Einarsson and to Hrafn hafnarlykill, both of them famous Vikings; to Hella-Björn, another 'great Viking', and to Thorbjörn the Bitter; to Ævar, who 'gave up Viking cruises to go out to Iceland' with all his sons, excepting Véfríd, 'who remained behind in the Viking way of life'.[6] There were Viking chieftains from the western seaboard of Norway, like Geirmund the Swarthy; and there were Viking bands which (after the battle of Hafrsfjörd) had established themselves in the lands *vestan um haf*, like those of Thránd and Sæmund.[7]

The route followed by the emigrants to the havens in the west of Iceland was by way of the south coast; and so round Reykjanes and up the west coast, with the north-going current, to Borgarfjörd, Breidafjörd, and the Western Fjords. But in the *Landnámabók* and the *Vatnsdæla saga* it is stated that Ingimund and his crew left Norway and reached Iceland, 'and had sailed the northern coast and then west, a thing never done before. Ingimund brought his ship into Hunavatnsós, and he gave all those place-names which have lasted ever since.' Later it is recorded of a certain crew, 'They voyaged to Iceland and sailed north-about round the island and stood westward to Sléttufjörd.' 'Grím . . . went to Iceland to seek for land there and sailed round the north coast.' Sometimes, as in the case of Aud the Wise, the route followed by the emigrants was by way of the Scottish isles and the Faeroes. On other occasions it is clear that the course lay across the open sea. From the time of the Settlement the dangerous south coast of Iceland was dreaded by mariners and was the scene of many wrecks; in the *Óláfs saga Tryggvasonar* it is said that there was 'nothing but sands and vast deserts and a harbourless coast, and, outside the skerries, a heavy surf'.[8] For this reason the south coast was settled last of all.

A fairly high proportion of the early voyages to Iceland appear to have been prosperous and uneventful. There is no suggestion anywhere in the *Landnámabók* that this venture was regarded as a noticeably difficult or dangerous one. On the other hand, mention is not infrequently made of long passages during which vessels were tossed for weeks and months on the high sea, and of voyages which ended in shipwreck.[9]

The immediate occasion for the departure from Norway of a good many chieftains was the battle fought between the rival fleets of Harald Hárfagri and the coalesced kings of Hordaland and Rogaland, about 892, at Hafrsfjörd.[10] Following upon this decisive victory over his enemies, Harald had made himself master of the entire country; and many of the chieftains opposed to his rule presently left the homeland to live as Vikings in the lands *vestan um haf*, returning in the summer to harry the coasts of Norway. Later King Harald took a fleet across the North Sea and evicted his enemies from the Scottish isles. Thránd, Sæmund, and a host of others sailed westward and established themselves in Iceland. It was in the years following the battle of Hafrsfjörd that a significantly high proportion of the emigrants left Norway for Iceland. Some of them were the sons of chieftains slain by the king's command; some, who failed to make good their escape, were also put to death; others, who had fought on Harald's side at Hafrsfjörd, emigrated on the king's advice.[11] But it can hardly be over-emphasized that the main and decisive factor in this great movement was the remarkable progress achieved during the eighth and ninth centuries in Norse shipbuilding, seamanship, and navigation.

By the reign of Harald Hárfagri in the latter part of the ninth century the Norsemen had evolved a strong ocean-going craft, variously known as the *hafskip*, *kaupskip*, and *knörr*, which was capable of transporting as many as three or four dozen men, a stock of cattle, the necessary food and fodder, and the furniture of a farmstead, several hundred miles across the ocean.

The *hafskip* appears to have developed on parallel lines during the same period that witnessed the evolution of the *langskip*. The *hafskip* was somewhat shorter than the *langskip*; it drew more water;[12] it was broader in the beam and with a much higher freeboard:[13] in strong winds it was a faster sailer.[14] Like the *langskip*, the *hafskip* was clincher-built and double-ended; and it was steered by a side-rudder secured to the starboard quarter.[15] The vessel was propelled by one large square-sail spread on a yard hoisted to a mast stepped amidships. The yard was hoisted and lowered by a halyard which also served as a backstay. The yard could be turned on its vertical axis by means of braces, *aktaumar*. Both tackle and anchor-cable were usually made of walrus-hide (*svarðreip*) on account of its great natural strength. When it became necessary to reduce sail the yard was lowered and one or more reefs were taken in; after which the yard was hauled up again. Little is known about the sails used in these vessels, except that each type of *hafskip* had its own distinctive sail, and that a well-cut sail was then (as now) held in great esteem among mariners.[16] The mast, consisting of a single pole of spruce or fir, was supported by shrouds, two of which were fastened to the gunnel on either side, and by a forestay, leading from the masthead to the prow. At the masthead was fixed a small pennon or wind-vane. Experience seems to have shown that ships of this type sailed better without any elaborate rigging. The lightness and simplicity of the ship's tackle were in keeping with the light and flexible 'structure of the ship's hull.

The cargo (*búlki*), covered over with ox-hides—on top of which was lashed the ship's boat—was stowed in an open hold amidships. The look-out was kept from the *búlkabrún*, or edge of the cargo-stack. When conditions were favourable the ship's tent or awning was rigged at night. The oars, which were few in number, were fitted fore and aft; they were commonly used for getting the vessel in and out of port,[17] and for assisting her to go about: they would be useless on a lee shore,[18] or in bad weather.[19] The *knörr* was essentially a sailing-vessel.

It is not generally realized that the *hafskip* of the Viking age was not much smaller, if at all, than the 'two Barkes' with which John Davis, in 1585, went in search of the North-West Passage. The carrying capacity of some of the *knerrir* used in the emigration to Iceland must clearly have been considerable. In any case a 40- or 50-ton trading-ship of the ninth century was fully as seaworthy as many later vessels of far greater tonnage.[20] In the thirteenth and fourteenth centuries there were ships sailing between Norway and Iceland that must have been of quite substantial size; complements of 60, 70, and 'nearly 90' persons are recorded.

With its light, flexible construction the *knörr* would rise easily to the scend of the heavy western sea; instances of foundering in the open Atlantic are rare; and the sagas relate how it rode out many a gale. The danger of *sagging* (i.e. compressive stresses amidships when the vessel was posed between the two wave-crests) was inconsiderable in view of the relative shortness of the *hafskip*; while the risk of *hogging* (i.e. when the stresses are reversed) was comparatively slight, since the cargo on board the *hafskip* was always carried amidships.

As has already been said, the earliest type of *knörr* was probably evolved some time before the close of the eighth century, during the great formative era of the *vestrvíking*, when the Norsemen were already making long ventures in the Atlantic to the Hebrides, Ireland, and the Faeroe Islands. The *knörr* was *par excellence* the normal ocean-going sailing ship of the Norsemen until well on in the Middle Ages. It was in a *knörr*, according to the *Landnámabók*, that Queen Aud voyaged to the Orkneys, Faeroes, and the Breidafjörd. The merchant Thórir went out to Iceland in a *knörr* which he had had built in the forests of Sogn. A *knörr* was also used by Grím and Kveld-Úlf for the passage to Iceland. It is significant, too, that some of the oldest place-names in Iceland were named after the *knörr*.[21]

By about 900, when the emigration was at its height, a considerable number of these merchantmen must have been engaged in the traffic, for in some years the greater part of 2,000 persons are believed to have made the passage. The size of the *hafskip* tended to increase in later centuries; but even as early as the Settlement they can scarcely have been much smaller than 40 tons. When Thórólf Móstrarskegg sailed for Iceland he took with him his wife and children, a good many of his friends, his household gear, and a heavy cargo of timber. Many another *hafskip* was similarly laden with livestock and gear. The *Landnámabók* tells of the livestock which Helgi brought out to Iceland, besides his wife and children; and of a craft which 'arrived in the estuary of Kolbein's river, laden with livestock'. The ship in which Thórir Pigeon-Nose sailed for Iceland was likewise laden. Each of the two *knerrir* in which Grím and Kveld-Úlf made the voyage appears to have carried a cargo of timber for house-building, besides a crew of thirty men. The ship in which Aud set out for Iceland must also have been of considerable size. Apart from her family, friends, and thralls, it is stated that Aud took twenty freedmen with her; and this was presumably in addition to the usual cargo of timber for house-building, farm and household gear, and livestock.[22] It was with such craft as these that the great emigration to Iceland—an event altogether unprecedented in European history—was accomplished.

About thirty-five of the chieftains, it is explicitly stated in the *Landnámabók*, came to Iceland in their own ships. The expedition of Ketill-Hæng comprised two, that of Geirmund the Swarthy four, *hafskip*. When the three sons of Thórgeir Vestarsson made the passage, each sailed in his own craft. Sometimes two emigrants owned a ship jointly. Sometimes two ships sailed in company. As in later centuries, ships were freely bought and sold. In many cases we know the name of the master, and sometimes the name of the ship is given. Mention is made of vessels which were renowned as fast sailers. Such a craft was the one King Harald presented to Ingimund—*Stígandi*, the *Flyer*, of which it was said that she strode over the seas.[23]

At the same time there had been a corresponding advance in Norse seamanship and navigation. Behind the plundering raids and ocean ventures of the Viking age lay uncounted centuries of seafaring experience. The *Landnámabók* gives testimony of the remarkably high standard of knowledge and proficiency displayed by the Norse mariner at the time of the Settlement.[24] *Vedrkœnn*, lying-to under some headland or island,[25] he watched the sky, the wind, the flight of birds, and the appearance of the coast around him; he would show much the same skill and caution when approaching the land as may be observed in the thirteenth-century sagas;[26] he would struggle for weeks against contrary winds;[27] in urgent need of water he would spread the ship's awning to catch the life-preserving rain;[28] caught on a lee shore, he would sometimes save the lives of the crew by running the ship

a.

b.

The coast and islands of Iceland
a.   The south coast and the Mýrdalsjökull
b.   The Breidarfjord
c.   The Vestmannaeyjar
d.   The Hornstrandir

ashore under full sail.[29] Already, with the aid of the tacking-spar, the *beitiáss*, he was able to beat to windward.[30] Four hundred years and more before the mariner's compass was known to the Norseman he was plying to and fro across the Norwegian Sea, the only mariner in Europe to venture so far and so often out of sight of the land. Judging from various significant passages in the *Landnámabók*, he was accustomed in this navigation to Iceland to rely upon a combination of dead reckoning and parallel sailing.[31] There are grounds, too, for believing that he may even have got some approximate idea of his northing or southing from a crude observation of the heavenly bodies.[32] Already, deprived of the sight of sun or stars by the heavy, overcast skies of the high North, he knew what it was to be adrift and hopelessly lost on the face of the waters: the plight of Örlyg and his men, as recorded in the *Landnámabók*, is perhaps the earliest known instance of *hafvilla*.[33] Already he was becoming increasingly familiar with the banks, deeps, rocks, and currents off the Iceland coast and with the seabirds, whales, and other adventitious aids to navigation in those waters.

For the best part of two generations the great exodus continued. The scale of the emigration, no less than the distances involved, was something new and astonishing. The *Landnámabók* has preserved the names of more than 400 chieftains of higher and lower degree, who, with their wives and families, friends, followers, and thralls, their livestock and farm and household goods, were transported 600 miles and more across the open sea in the small, sturdy ocean-going sailing craft which, evolved something like a century earlier, had made possible the overseas expansion of the Viking age. From the fertile vales of the Trondelag, from the wooded fells of Hordaland and Rogaland, from the fields and forests of Agder and the Vestfold, came peasant-farmers and fishers in their thousands bent on emigration. Some of these made the voyage of their own free will; others sailed away to escape from danger or oppression in the homeland. Some of the emigrants were well stricken in years, and a number of them died at sea. Vessel after vessel was crammed full of human and animal freight, lay-to for a while off the land, and then squared away, before a favouring breeze, for the distant island across the ocean.

After a voyage of varying duration the coast of Iceland would at last slowly lift above the horizon; the snow-capped mountains and vast glimmering glaciers would imperceptibly grow more and more distinct; the surf-lashed beaches, undulating meadowlands, and extensive birch-woods which at that time flourished on many parts of the coast would gradually become visible; certain well-remembered landmarks—the gigantic *massif* of the Mýrdalsjökull, the columned cliffs of the Vestmannaeyjar, the headlands of Langanes, Reykjanes, and Snae-fellsnes—would help to guide them to their destined havens.

Some of the ships would stand down the long, featureless strand of the harbourless south coast, and, rounding the land, pass up into the broad, shining waters of Faxafloi and on to the Breidafjörd with its myriad islands. Presently, while the *knerrir* lay at their moorings, the cows, sheep, goats, and horses of the settlers would be disembarked and driven off to the virgin pastures; the timber and other gear would be carried ashore, and the first makeshift dwellings erected. Some of the boldest seafarers, many of the most distinguished chieftains, a substantial pro-portion of the most vigorous stock in Norway and the lands *vestan um haf*, went to swell the steady stream of emigrants. The descendants of these settlers inhabit the country to this day.

9

The explorations of Eirík Raudi

THE FIRST RECORDED allusion to Greenland is to be found in the *Landnámabók*. According to Icelandic tradition, a mariner called Gunnbjörn Úlfsson was, early in the tenth century, while on passage to that island, driven far to the westward by foul weather to a group of islets which henceforth bore the name of Gunnbjarnarsker, or Gunnbjörn's Skerries:[1] from which point, on the western horizon, he had described an unknown shore. Gunnbjörn apparently made no attempt to reach this shore; but, as soon as he got a fair wind, sailed back to Iceland, where his sons and brother were already settled.

This, at any rate, is the generally accepted view as to the first sighting of the Greenland coast by the Icelanders: but there exist other possibilities. It was suggested in an earlier chapter that the Arctic mirage, a phenomenon sufficiently common in high latitudes, was perhaps a significant factor in exploration and discovery.[2] This could have been the case as regards Greenland. The Icelanders may, in fact, have learned of its existence through a mirage. Or again, it is possible that they drew the logical conclusion from observing the flight-line of wild geese or other migratory birds passing between the Faeroe Islands and Greenland.

Apparently, therefore, it was either the report of Gunnbjörn's exploit, or knowledge derived from the flight-line of migrant birds and/or the Arctic mirage (or *hillingar* effect), that had convinced Icelanders that there must be land somewhere away to the westward. In any event, they presently proceeded to act as if they had full and certain knowledge to that effect.

Gunnbjörn's sons were living on the Ísafjörd in the West Firths when a new settler and his family arrived in Iceland from Norway. His name was Thorvald Asvaldsson, and he came from the Jaeder, near Stavanger. His son, later renowned as Eirík Raudi, was about sixteen when with his parents he had to leave the country on account of a feud and a homicide. They settled in the West Firths at Drangar, on the bleak and inhospitable coast of Hornstrandir; this was the best they could do, since by this time nearly all the good land in Iceland had been taken. The story of Gunnbjörn and his skerries must have been well known to Eirík from his boyhood.[3]

When he was a grown man and married Eirík left Drangar for Haukadal, where there was good grazing. There he was involved in quarrels and blood-feuds with his more powerful neighbours and presently driven out of Haukadal. He took refuge on one of the islands in the entrance to the Hvamsfjörd. Again there was a quarrel, in the course of which Eirík slew two of his enemy's sons. This was followed by a sentence of lesser outlawry; Eirík being banished overseas for the term of three years.

In spite of these reverses Eirík's friends continued to assist him and successfully concealed him from his enemies while a ship was being fitted out in the Breidafjörd to carry him from Iceland. Eirík declared his intention of going in search of the land which Gunnbjörn Úlfsson had sighted many years before. He would return and see them again, he told his friends, should he discover that land. After a while he got a fair wind and sailed out of the Breidafjörd. Taking his departure from Snaefellsnes, he steered due west across the ocean.

On sighting the skerries which had been named after Gunnbjörn (these are generally believed to have been a group of rock-islets lying off the East Greenland coast, approximately midway on the track between Snaefellsnes and the Eiríksfjörd), Eirík sailed on and made his landfall at a point on the East Greenland coast which he called Midjökull (Middle Glacier). It was a barren, gaunt, forbidding shore of lofty mountains and precipitous cliffs and headlands, broken by fjords of varying breadths, with a glint of glaciers beyond, and fringed by rock-islands and wave-lashed skerries—utterly uninhabitable. And away beyond the frowning cliffs and mountains, white and gleaming in the sun, was the everlasting ice-cap.

They stood southward with the current down the coast, rounded Hvarf (now Cape Farewell), and put in at that district which was to become known in after years as the East Settlement, Eystribygd. This was an altogether different matter from the inhospitable east coast. Here were deep channels and safe anchorages. A proper sailor's coast. Higher up the fjords were fine pastures, which held out good prospects of stock-raising. A farmer could make something of this land. It might be possible to establish a settlement here, as their countrymen had already settled Iceland. Above all, there were no hostile neighbours to harass the fiery Eirík, as there had been in Haukdal and the Breidafjörd. If he took possession of this land and succeeded in persuading men to follow him here, he would indeed be monarch of all he surveyed.

Eirík passed his first winter on an island he named Eiríksey, 'Eirík's Island', and next spring sailed up the Eiríksfjörd to a fertile inland plain, where he built a house. In the summer he continued his exploration of the new country, bestowing place-names far and wide. Nearly two hundred miles to the northward was another network of fjords penetrating deep into the interior, with lush grassy slopes along their shores: though beyond, as everywhere else on this coast, were high mountains and the ice-cap. Only a comparatively narrow coastal strip on the south-west coast of the new land was fit for habitation.

They spent the second winter at Holmar among the islands near Hvarf, the southernmost point of Greenland. The following summer Eirík sailed to the northward, this time probably up the east coast (where he may have come upon traces of former inhabitants, the Eskimos), as far north as the parallel of the head of the Eiríksfjörd. He afterwards returned and wintered again at Eiríksey.

The term of Eirík's banisment from Iceland was now over. The three years were up and he was free to return to the island. That summer he stood down the west coast, rounded Hvarf, and sailed back across that part of the ocean afterwards known as the Greenland Sea, steering for the Snaefellsjökull and bringing his ship safely into Breidafjörd.

Soon after Eirík's return, the blood-feud with his formidable neighbours blazed up anew. It was evident that his days in Iceland were numbered. The only possible future for a man such as he lay over the ocean. This was Eirík's last winter in Iceland. He called the country which he had discovered 'Greenland', because, he said, he thought that giving the place a good name would induce men to go there. The event proved him right. When in the following summer Eirík sailed westward again to settle in Greenland for good, he was accompanied—or shortly after was joined out there—by a number of staunch friends who had stood by him in his hour of need and who were now to reap their reward in the shape of lavish grants of land in the newly discovered country. Among these friends were Thorbjörn Vifilson from Langabrekka, 'who sold his land and bought a ship and sailed with thirty men to Greenland', accompanied by his beautiful young daughter, Gudrid, one of the most attractive characters in Norse history; and Herjólf Bárdarson, son of Bárd Herjólfsson from Sognfjörd, a relative of Ingólf, the first Norseman to settle in Iceland, who came to dwell at Herjólfsnes near Hvarf.

In the course of the next few years Greenland was colonized from Iceland.[4] Once again a fleet of 'floating Noah's arks' passed overseas to settle in the empty lands. 'When Eirík returned to Iceland he "sold" the new country so attractively to his neighbours that he was accompanied back to Greenland in 985 or 986 by 14 ships carrying between 400 and 500 persons and with them horses, cattle, sheep, goats, chickens, dogs, cats, and varied household goods.'[5] The emigration to Greenland was therefore very like the earlier settlement of Iceland, though of course on a much smaller scale. These ventures were destined to mark the final stage of Norse expansion to the westward.

They had about 450 miles to sail. By this time they must have had some rudimentary knowledge of the navigation of the Greenland Sea. The warm Irminger Current sets northward along the west coast of Iceland. In the vicinity of the 100-fathom line there is a well-defined line of demarcation between this warm north-flowing water and the cold south-flowing water of the East Greenland Current which occupies most of the Denmark Strait. At this stage of the voyage the emigrants would in all likelihood become aware of the pronounced change in the colour and temperature of the water and a no less significant change in the local fauna. Once in the south-flowing Polar current the ships would be set southward down the East Greenland coast, around Hvarf, and then northward to the habitable regions of the great island.

It is recorded in the Landnámabók that, according to Ari Fródi, besides the fourteen ships which arrived safely in Greenland, there were eleven others that, sailing from the Borgarfjörd and Breidafjörd at this time, were either driven back again or lost.[6] Nothing is known about the fate of the vessels which were lost. They may have foundered in the open sea, perhaps engulfed by the dread hafgerðingar which Herjólf Bárdarson encountered;[7] or the long, heavy swell rolling landwards may have swept them on to the ice and skerries of the East Greenland coast.

Some doubt has been cast upon Ari Fródi's testimony to the effect that twenty-five ships sailed from Iceland for Greenland with the colonists and their belongings. It is claimed by Lúðvik Kristjánsson that the greater part of this emigrant fleet consisted, not of ships (i.e. hafskip), but of open boats similar to the 'ten-oared boats' of the Saga age used for the fishing and coastal traffic, which, it is further claimed, were really more suitable for the passage than the knörr. 'There is good reason to believe that it was the fault of the *knerrir*, rather than that of the "ten-oarings", that 44 per cent of Eirík's fleet was either lost or forced to turn back.' It is argued that the Borgarfjörd and the Breidafjörd, from which the emigrant fleet sailed, could not possibly have furnished as many as twenty-five *knerrir*.[8]

The essential weakness of this argument is that it altogether leaves out of account the realities of the 'sea affair'. Can the writer really believe (as he says) that the ships of the Saga age took no more than *four days* to sail from Snaefellsnes to Hvarf? Does he think that a rowing-boat, however stoutly built, was actually more seaworthy than the *knörr*? The fact is, there is nothing like real evidence in support of these suggestions. It could very well be months before a ship reached a Greenland haven from Iceland: if the crew arrived at their destination in a matter of weeks, they were lucky. Generations of experience had gone to show that the *knörr* was well fitted for the navigation of the North Atlantic. No other type of vessel is ever recorded to have made the voyage to Greenland. In the last two decades of the tenth century the Icelanders certainly owned a good many *knerrir*, or *hafskip*.[9] With so many chieftains resolved to follow Eirík to Greenland such vessels must have been at a premium: their owners had every inducement to sell. This argument against the *knörr* creates more problems than it solves. On the basis of all the available evidence, therefore, there are no convincing grounds for asserting that the Greenland fleet was composed of a different type of craft from that which had been used in the great emigration to Iceland.

The discovery of Greenland by the Norsemen followed a pattern which was already familiar—first, the fortuitous sighting of a distant coast by some storm-tossed ship's company; next, a regular, organized voyage of exploration; and, finally, the passage of considerable numbers of emigrants across the sea to take possession of the new land. It is worth noticing, too, that the earliest and most authoritative source, Ari Fródi, clearly ascribes to Eirík Raudi the honour of being the first to land and settle in Greenland, and that the settlement began in about 985. There is no question that this formidable red-haired, red-bearded chieftain was the father of the new colony and that the history of Greenland stems from him.

The country which is called Greenland was discovered and settled from Iceland. Eirík Raudi was the name of a man from Breidafjörd who went out there from here and took possession of land in the place which has ever since been called Eiríksfjörd. He gave the country a name and called it Greenland, because he said that men would want to go there if the land had a good name. . . . He began to colonize the country fourteen or fifteen winters before Christianity came to Iceland, according to what a man who had gone out there with Eirík Raudi told Thorkel Gellison in Greenland.[10]

10

The voyage of Bjarni Herjólfsson

JUST AS THE discovery of Greenland was a natural, almost inevitable, consequence of the Norse navigation to Iceland, so the first sighting of various parts of the North American seaboard was the outcome of the new traffic with Greenland. According to the *Grœnlendinga saga*, the honour of steering the first European keel into American coastal waters belongs to one Bjarni Herjólfsson.

It is now generally recognized that this narration, and not, as was formerly supposed, the *Eiríks saga rauda*, is the older and more dependable source for the various Norse voyages to America, and that it preserves the original and authentic account of their discovery of the new lands across the ocean. On the evidence of the genealogies, and for certain other strong and compelling reasons, the late Professor Jón Jóhannesson in a lucid and masterly paper published in 1956 ascribed the *Grœnlendinga saga* to the last part of the twelfth century.[1] The *Grœnlendinga saga* is, therefore, one of the oldest of the Icelandic sagas. Jóhannesson showed that Bjarni Herjólfsson, and not Leif Eiríksson, was the first discoverer of America; that it was Thorsteinn, and not Leif Eiríksson, who was 'tossed on the high sea for a whole summer'; that the *Grœnlendinga saga*, and not the *Eiríks saga rauda*, preserves the more trustworthy narrative of the voyages to Vinland; that the author of the *Eiríks saga rauda* drew upon the *Grœnlendinga saga* for much of his material, adapting it to his own purposes; that the tradition of the voyages made by Karlsefni and the others to Vinland has come down through Karlsefni's family; and that the *Grœnlendinga saga* dates back to the time of Brand Sæmundarson, bishop of Hólar (who may, indeed, have written the saga himself), the great-grandson of Snorri Thorfinnsson, who was the first white child to be born in America.

It is many years since G. M. Gathorne-Hardy first drew attention to the fact that the account of Bjarni Herjólfsson's voyage was full of nautical phraseology and details of interest only to sailors; such as getting one's bearings from a view of the sun; the sighting, and losing sight, of the land; the hoisting, taking in, and reefing, of sail; standing in to, and standing off from, the land.[2] This is an interesting and significant point; for the narration, if true, must necessarily have originated with Bjarni himself or else with one of his crew. And it has been handed down to posterity by another remarkable Icelandic seaman, Thorfinn Karlsefni.

'And of all men', the *Grœnlendinga saga* concludes, 'Karlsefni has given the most complete account of all these voyages, the account of which has to some extent been related here.'[3] The saga itself gives ample testimony of its origins; they are, in fact, to be traced back to Karlsefni and his wife, Gudrid; and the tradition has come down to us through their descendants.

The following account of the voyage of Bjarni Herjólfsson, who may in truth be considered the first discoverer of America, is therefore based on the *Grænlendinga saga*.[4]

Among the early settlers of Iceland was Bárd Herjólfsson. His son, Herjólf Bárdarson, had a son called Bjarni, a most promising young man. Bjarni had been a seafarer from his youth. His voyages had brought him much wealth and renown. He was accustomed to pass his winters alternately overseas and in Iceland with his parents. Before long he was able to purchase a merchant ship of his own. One winter when Bjarni was away in Norway, his father made up his mind to accompany Eirík Raudi to Greenland and make his home there. It is said that on this voyage Herjólf had on board with him a Christian man from the Sudreyjar who composed some verses about the dreaded *hafgerðingar*, or 'sea-hedges', from which they appear to have had a narrow escape. As has already been said, Herjólf made his home in Greenland at the place which was afterwards known as Herjólfsnes, in the vicinity of Hvarf, and was much respected in the colony.

Meanwhile Bjarni had returned to Iceland; but on his arrival at Eyrarbakki it was only to learn that Herjólf had sold his lands in the island and sailed for Greenland. Bjarni was much troubled by the news of his father's departure and refused to unload his cargo. When his crew wanted to know what he intended to do, he told them that he meant to follow his usual practice, after his absence in Norway, of passing the next winter with his parents.

'I mean,' said he, 'to sail my ship to Greenland, if you are willing to accompany me.'

Thereupon the whole crew declared that they would stand by this decision, and Bjarni continued:

'Our voyage will appear rash, seeing that none of us has experience of the Greenland Sea. . . .'

In spite of this Bjarni and the others set off as soon as they had made their ship ready for sea and sailed for three days[5] before they lost sight of the land; but after that the fair wind died away, and northerly winds and fogs succeeded, and they did not know which way they were going; and this went on for many days.

At last they saw the sun and were able to get their bearings. Thereupon they made sail; and after sailing that day they sighted land, and argued among themselves what land it might be. Bjarni, however, said he did not think it could be Greenland. The others asked him whether he meant to stand in to the land or not.

'I am going', said Bjarni, 'to sail close in with the land.'

This they did, and it was soon apparent that the land was not mountainous and was thickly wooded, with a number of low hills. They stood out to sea again. Bjarni's one idea was to shape course for their original objective, Greenland. The unknown shore he saw across the water had no attraction for him. His companions and he had only just emerged from the state of *hafvilla*, which implied loss of both reckoning and sense of direction: an unnerving experience for the most seasoned mariner. Moreover, the visible change in the altitude of the midday sun or the Pole-star at night may well have warned him that he was a long way too far to the southward.

After sailing for two days they saw another land. Again the others asked Bjarni whether he thought this was Greenland. Bjarni replied that he did not believe this was Greenland any more than the other land had been.

'For', said he, 'it is reported that there are very large glaciers in Greenland.'
They soon approached this land and saw that the country was low lying and well wooded. At this point the fair wind died away altogether, and the crew suggested that it would be a good thing to land there; but Bjarni would have none of it. The men then grumbled that they were in want of both food and water.

'You are in no need of either,' retorted Bjarni, to the loud and vociferous complaints of his crew.

Bjarni ordered them to make sail again, and this they did; and then they stood out to sea and sailed for three days before a south-westerly wind, after which they

Page from the Flateyjarbók (p. 223) recording the first European sighting of America

sighted a third land. They saw that the land was high and mountainous, and there was ice. Again the crew asked Bjarni if they should land there; and again he refused, observing that the land appeared to him to be good for nothing.

According to Bjarni's reckoning, they were still too far to the southward for that coast to be Greenland. So, without lowering the sail they held on their course along the land, and saw that it was an island. Once more they stood out to sea, and sailed on with the same wind; but the wind freshened, and Bjarni ordered them to reef, and not crowd more sail than the ship and rigging would stand. They sailed for four days, and then they saw a fourth land. The crew asked Bjarni if he thought this was Greenland. This time he replied that he indeed believed it was.

'This,' observed Bjarni, 'is most like what I have been told of Greenland, and here we will stand in for the land.'

They did so, and on the evening of the same day they drew in under a certain headland, where they saw a boat. It was in this place that Bjarni's father had made his home. The headland was named after him and was known for many years to come as Herjólfsnes.

Bjarni went to his father's home and gave up his seafaring life; and lived there for the rest of his father's lifetime, afterwards inheriting his estate.[6]

From the geographical point of view, no doubt the sighting of the American coast for the first time in history by a shipload of Norse mariners is of prime importance. To the maritime historian, however, the navigational aspect of the voyage is of even greater interest and significance. After completely losing their reckoning, when they had had no idea at all where they were or which way they were heading, they had sighted this unknown coast. Nevertheless, Bjarni extricated himself and his companions from the state of *hafvilla* and achieved his purpose; he continued his voyage, and in due course arrived safely at his destination—Greenland. 'That he did so with speed and precision,' observes Gathorne-Hardy, 'might give cause for surprise, were there not many well-authenticated instances in Icelandic literature of men who, after drifting about, the sport of adverse winds and fogs for a long time, retained to the last sufficient knowledge of their position to enable them to return home.'[7]

This scarcely does justice to Bjarni's achievement. It is again to be emphasized that he did *not* retain sufficient knowledge of his position to enable him to lay a course—for it is plainly implied in the saga that he had no such knowledge; all on board are stated to have been in the state of *hafvilla*.[8] What he and others in like case did achieve was, in fact, something far more remarkable and significant: namely, after having lost their reckoning and all knowledge of their position, they were nevertheless able once again to shape the right course for their destination. This, as will later be seen, goes to the very root of the matter and is the key to the proper understanding of the Norse methods of ocean navigation in these early ventures.

Like John Cabot several hundred years later, Bjarni had had trouble with his crew. He would not permit that to deflect him from his purpose. He was not deceived by the apparent likeness of the penultimate land they sighted to what they had been told about Greenland. He still held on his northerly course. In spite of appearances, Bjarni trusted to his navigation, and the others in the ship, despite their grumbling, acquiesced.

11

The settlement of Greenland

ALONG THE MOUNTAINOUS, deeply indented seaboard north of Cape Farewell—so strangely like parts of the west coast of Norway—the settlers made their homes. The climate appears to have been then much milder than in later years. Some of the farmsteads were built on or near fjords that today are blocked with ice. There was good living in Greenland. In some respects conditions compared favourably with those in Iceland. Near the head of the fjords were wide meadows which supported large herds of cattle and sheep. On the dry uplands there was extensive sheep-grazing. Both the sea and the rivers abounded in fish. The country, too, was rich in game.

Eirík Raudi fixed his dwelling-place at Brattahlid, at the head of the Eiríksfjörd. The lesser chieftains each took possession of a fjord and in most cases named it after himself. Thus Herjólf took Herjólfsfjörd; Ketil took Ketilsfjörd; Hrafn took Hrafnsfjörd; Solvi took Solvadal; Helgi Thorbrandsson took Alptafjörd; Thorbjörn Glora took Siglufjörd; Einar took Einarsfjörd; Hafgrím took Hafgrímsfjörd and Vatnahverfi; and Arnlaug took Arnlaugsfjörd. The franklins and other folk also established themselves throughout the country. Soon they called themselves Greenlanders. The descendants of these settlers occupied the land for upwards of five hundred years.

Brattahlid was the centre of the Eystribygd (East Settlement). The Vestribygd (West Settlement) lay about 180 miles higher up the coast. Approximately midway between the Eystribygd and the Vestribygd was a smaller settlement, which is called the Middle Settlement, about which little is known.[1] The average farmstead of the West Settlement was smaller than that of the East Settlement. The *Thing*, or assembly for the whole colony, was held near Gardar. The objects which the settlers had in view when choosing a site were a sheltered cove, a good stream, and a fairly level space around the farmstead for the 'homefield'. The Greenlanders wisely preferred the north side of the valleys, or the 'sun side', for their dwelling-places.

In the anonymous *Konungs Skuggsjá*, *The King's Mirror*, written about the year 1250, there is a detailed and most interesting account of conditions in Greenland. From this work, which takes the form of a dialogue between father and son, a vivid and in certain respects remarkably accurate impression is conveyed of the everyday life of the Greenlanders.

In reply to his son the father says:

You asked whether the sun shines in Greenland and whether there ever happens to be fair weather there as in other countries; and you shall know of a truth that the land has beautiful sunshine and is said to have a rather pleasant climate. The sun's course varies greatly, here; when winter is on, the night is almost continuous; but when it is summer, there is almost continuous daylight. When the sun rises highest, it has abundant power to shine and give light, but very little to give warmth and heat; still, it has sufficient strength, where the ground is free from ice, to warm the soil so that the earth yields good and fragrant grass. . . . The people in that country are few, for only a small part is sufficiently free from ice to be habitable; but the people are all Christians and have churches and priests. If the land lay near to some other country it might be reckoned a third of a bishopric; but the Greenlanders now have their own bishop, as no other arrangement is possible on account of the great distance from other people.[2]

According to Konungs Skuggsjá, the pasturage was good, and there were fine and large farms in Greenland.

The farmers raise cattle and sheep in large numbers and make butter and cheese in great quantities. The people subsist chiefly on these foods and on beef; but they also eat the flesh of various kinds of game, such as reindeer, whales, seals, and bears. That is what men live on in that country.[3]

It is stated that in Greenland there were reindeer in great abundance, and that hares and wolves were very plentiful.

There are bears, too, in that region; they are white, and people think they are native to the country, for they differ very much in their habits from the black bears that roam the forests. These kill horses, cattle, and other beasts to feed upon; but the white bear of Greenland wanders most of the time about on the ice in the sea, hunting seals and whales and feeding upon them. It is also as skilful a swimmer as any seal or whale. . . .[4]

The testimony of the Konungs Skuggsjá is borne out by the evidence of archaeology. In the East and West Settlements the remains of nearly 300 steadings have been discovered. The most populous region of the colony was the fertile vale of Gardar, lying between two broad fjords and sheltered from rough weather by the surrounding high mountains, and also the wide, grassy hill-slopes of Vatnahverfi, where there are the ruins of at least forty farms, most of them far inland, and good grazing and reindeer herds. The number of cows kept by the Greenlanders was remarkably large. A middle-sized farm in the East Settlement would support as many as between ten and twenty head of cattle. Their flocks of sheep were also considerable and the Greenlanders produced much homespun cloth. It has been conjectured that in its most flourishing period the total population must have amounted to something like 3,500 souls. The figure should, perhaps, be even higher.

Most of these early settlers in Greenland, as has been said, came from Iceland. There were also a small number of Norwegians. After the see of Gardar was set up in the twelfth century, successive bishops of Greenland probably took out a few followers with them to their distant diocese. It is also possible that the various merchants and mariners wintering in Greenland occasionally introduced new blood into the colony.

For a number of years the course for Greenland followed the track taken by Eirík Raudi on the original voyage: namely, due west across the sea from Snaefellsnes to the East Greenland coast off Midjökull—a first-rate mark—and then southward down the coast and around Hvarf, after which approach could be made to the East Settlement. It has been stated above that an interesting point about these early voyages is that ice conditions appear to have been a good deal less severe than was

Greenland: Gardar (above) and Brattahlid (below)

the case in later centuries. An early departure such as Eirík's, in the spring, would be out of the question today owing to the vast quantities of ice in the Greenland Sea. The absence of impassable ice must have greatly assisted the colonization of Greenland, which was completed in little more than a generation: it occupied the years approximately between 985 and 1020.

The year 999 marks an era in the annals of the sea. For in that year Leif, son of Eirík Raudi, abandoning the old course by way of Midjökull and Snaefellsnes, steered straight across the Atlantic for Norway. The following summer he sailed back across the ocean. This was the beginning of direct traffic between Greenland and Europe, and the first transatlantic trade-route in history. This (and not the first discovery of America, which was for long erroneously attributed to him) was Leif Eiríksson's outstanding achievement. The course which Leif followed is unknown; but there is reason to believe that, at any rate in later years, the Norwegians usually crossed the Western Ocean on the parallel of Hernar, an island-group on the west coast of Norway some thirty miles north of Bergen. It was a track which lay well to the southward of the dangerous south coast of Iceland, on which so many craft were to meet their end. That it involved the crossing of a vast and stormy region of the Western Ocean would seem to be conclusive evidence of the high standard of seamanship and navigation attained by the Norsemen in these early times.

The transatlantic voyages thus begun continued for more than four centuries. The significance of the Greenland passage has to some extent been obscured by the superior interest, in the public eye, of the Norse discovery of America. Compared with the considerable volume of literature which has been poured out on that subject during the last hundred years and more very little has appeared concerning this first transatlantic trade-route. Yet from the standpoint of economic and maritime history the long-continued traffic between Norway and Greenland is of far more consequence than the occasional voyages that the settlers in Greenland made to the North American littoral, which entailed no such crossing.

In the opening years of the eleventh century quite a number of the settlers were sailors as well as franklins. Leif Eiríksson and, somewhat later, Skúf in the *Fostbrœðra saga* were true deep-watermen, *hafsiglingarmenn*, sailing to and fro between Greenland and Europe. Most of the overseas traffic of Greenland was with Norway; but a good many voyages, particularly in these early years, were made to Iceland. There was at this time a fairly close connection between the two lands. (In ecclesiastical affairs Greenland was considered to be within the jurisdiction of the see of Skálaholt.) After about the year 1000 the Norsemen turned more and more from marauding to peaceful traffic. Trade and 'sailorizing' went hand in hand. The seafaring trader was regarded in both Norway and her colonies as an honoured visitor. In the later medieval period the Greenland trade was mainly in the hands of Norwegian merchants, who used to sail to Greenland during one summer and return to Norway in the next.

In the eleventh and early twelfth centuries, when Scandinavian maritime enterprise was at its height, we sometimes hear of several merchant ships entering the Eiríksfjörd in a single year. On arriving in the colony the merchants would set up their booths and trade with the settlers. Their ships would be laid up for the winter and the merchants received into the homes of the neighbouring farmers. It is worth noting that the cargoes carried were sufficiently substantial to require large outbuildings for their storage during the winter. Thus, when Skúf came out to

Greenland, about the year 1025, there was the customary gathering at the water's edge to exchange news, to examine the merchants' wares, and to do business. Skúf's ship entered the Eiríksfjörd late in the autumn. Thorkel Leifsson was chieftain at Brattahlid at the time. 'He went on board the ship as soon as it was moored and bought from Skúf and the other men all that he needed. He bought all the malt they had, and other things which were hard to come by in Greenland.' Skúf and his partner, Bjarni, owned a farm on Stockness, on the opposite side of the Eiríksfjörd to Brattahlid. Bjarni managed the farm while Skúf was away at sea. The chieftain's homestead at Brattahlid was crowded with guests that Yule. Skúf and Bjarni were among the guests invited by Thorkel. There was great feasting and drinking of home-brewed ale, and joy and merriment.[5]

The history of medieval Greenland might be fitly described as a series of illuminative glimpses of life and conditions in the colony, alternating with long periods of almost complete darkness. The earliest of these glimpses, as we have seen, is of Eirík Raudi and his times; the next, of Brattahlid under Thorkel Leifsson; and after that, of the Eystribygd in the early part of the twelfth century.

The colonists had been Christian, more or less, since the early years of the eleventh century. It is recorded in the *Eiríks saga rauða* that that unregenerate old pagan, Eirík Raudi, had stubbornly opposed the new teachings (it was because of one of Eirík's sour comments that Leif Eiríksson has gone down in history as the man who brought 'the charlatan'—i.e. the first Christian priest—to Greenland); but Leif and his wife, Thjodhild, had embraced the Faith, and Eirík's own wife actually left him, refusing to live any longer with a 'heathen man'. The tradition of Eirík's recalcitrance has recently been corroborated by the evidence of archaeology. The first Christian church at Brattahlid was built at a discreet distance from the homestead. It is not known whether or not Eirík was converted before his death. Later in the eleventh century it would seem from the *Gesta* of Adam of Bremen that it was difficult to procure sufficient priests to supply the needs of the Greenlanders.[6]

The first bishop ever to set foot in Greenland was an Icelander, Eirík Upsi. Little is known about Eirík or his precise status. It is at any rate certain that he was not bishop of Gardar, for the see had not been established. In 1112 Eirík sailed out to Greenland, and nine years later made one of the most mysterious disappearances in history. A line in the Icelandic annals briefly records his fate: 'Bishop Eirík went in search of Vinland'.[7]

Soon after this it was resolved to send emissaries to Europe to ask that a bishopric might be established in the land. The lead in this matter was taken by the chief man in Greenland, Sokki Thorisson of Brattahlid; and eventually the mission was entrusted to Sokki's son, Einar, who in 1123 sailed for Norway with a large cargo of ivory and walrus-hides, together with a live polar bear as a present for the king. Einar's mission was successful. On the return passage he was accompanied by the newly consecrated Bishop Arnald, to whom he had sworn an oath that he would 'uphold and maintain the rights of the bishop's see' and defend them against all comers.

The perils of the long ocean venture to Greenland are well illustrated by the events that followed. In the face of contrary winds, Einar at last gave up the attempt to reach the Eiríksfjörd that season and put into a haven on the south-west coast of Iceland, in which island the bishop and he spent the winter. But no news was received of another trading-ship, of which Arnbjörn the Norwegian was master, which had set out for Greenland at the same time. In the following summer (1126) the bishop and Einar continued their voyage and came safely to Greenland. At Gardar, on the long narrow peninsula between Eiríksfjörd and Einarsfjörd, the bishop established his see. On this site were later to arise the cathedral church and the episcopal residence.

Some time later a party of Greenlanders, led by Sigurd Njálsson, described as 'an experienced mariner', who was accustomed to go on long hunting and foraging expeditions into the uninhabited parts of the country, were debating whether or not to abandon an expedition which had so far proved disappointing and unrewarding. The summer being nearly over, Sigurd's shipmates were in favour of turning back. 'It was risky work,' they argued, 'sailing up these great fjords under the glaciers.' Nevertheless, Sigurd himself was inclined to press on further into the desert (úbygd), along the barren, succourless, ice-bound eastern seaboard of Greenland, because he had a presentiment that their expedition might not turn out to be unprofitable after all. The event proved Sigurd right. Shortly after, on sailing into a nearby fjord, they came suddenly on a merchant-ship beached in an estuary, and the camp which the unfortunate Arnbjörn and his companions had set up, after struggling ashore from their shipwrecked knörr. Arnbjörn and his companions now lay dead: and their possessions were fair salvage. The knörr, or trading-ship, being badly holed below the water-line, was damaged beyond repair; but Sigurd and the others bore away with them a smaller craft, 'a ship with a figure-head and painted', also the ship's boat. These, together with a large sum of money that they had discovered in the camp, they had brought back with them to Gardar. Sigurd proposed that the money and the 'figure-head ship' should be made over to the see for the good of the dead men's souls; and this was agreed to. The rest of the goods were shared among his party according to the law of Greenland.

When Arnbjörn's nephew Özur in Norway received news of these happenings he sailed for Greenland with other relatives of the deceased who expected a share in the inheritance. 'They [the merchants] got to the Eiríksfjörd from Norway and men came down to meet them and to do business. Then they were conducted to their quarters. Özur, the master, went to Gardar to the bishop, and was there throughout the winter.' It happened that there were two other vessels, one belonging to a Norwegian, Kolbein Thorljotsson, and the other to two Icelanders, Hermund and Thorgils Kodransson, in Greenland at the time, but both were away at the West Settlement. Özur, failing to carry his point with the bishop, quitted his host in deep dudgeon, and presently took his suit to the *Thing*, where he likewise failed to receive satisfaction. Feeling that he had been shamefully treated, he then deliberately and irreparably damaged the vessel which Sigurd had salvaged. Thereon the quarrel developed along lines familiar in Norse history. The bishop having declared Özur's life forfeit and called on Einar to take action, Özur was treacherously struck down by Einar at the Christmas feast. Some months later Kolbein and the other merchants (who had been prevented from attending the *Thing* by head-winds) arrived from the Vestribygd. At Gardar, as the cathedral bell was heard ringing for mass, Kolbein predicted darkly that it might well turn out that before the end of the day this could change to a funeral knell. A meeting was arranged between the Greenlanders and Norwegians in the hope of securing a peaceful settlement. However, at the meeting a quarrel broke out, followed by a murderous affray, in which a number of both the Norwegians and the Greenlanders were killed, including Einar, who, mortally wounded, breathed his last at the knees of his friend the bishop. Besides the dead, there were a good many others badly wounded. Kolbein and his followers got on board their ship and crossed the Einarsfjörd to the place where the other two *knerrir* were being hurriedly made ready for sea.

'You said truly, Kolbein,' said Ketil, 'that we should hear a funeral bell before our departure, and it is my belief that Einar's corpse is being carried into church.'

Kolbein, who in the course of the affray had slipped round behind Einar and felled him with his axe, confessed that he was in some degree to blame for what had occurred.

'We may expect the Greenlanders to launch an attack on us,' Ketil went on. 'So my advice is for our men to keep on loading as hard as they can, and for all of us to stay in the ship by night.'

The Greenlanders, as might be expected, had considerable doubts about their chances of success if it came to a fight between their own small craft and the large Norwegian *knerrir*. Shortly after the latter were ready to sail; but suddenly the ice swept in and the fjords were blocked. It appeared as if the Greenlanders would have their chance of revenge after all; then, at the top of the springs, the ice dispersed and the vessels got away; and the Norwegians came safely to Norway.[8]

After the lurid events related above, the curtain again descends on the Greenland scene. For rather more than a century little is known about that distant land. But it would appear that the colony throve and a more or less regular traffic was kept up with Norway. When a new archbishopric was established at Nidarós, or Trondheim, in 1152, Greenland as well as Iceland and the Faeroes became part of this archdiocese. New churches were added to those already built. The cathedral at Gardar dates from about 1200. Despite the remoteness and isolation of the colony it was none the less an integral part of Christendom.

In the East Settlement there were twelve, and in the West Settlement four, parish churches. The high proportion of churches to population is accounted for by the fact that many of the homesteads were scattered over an immense area. The best preserved of these ruined churches is at Hvalsey, near the modern Julianahaab; it dates back to the early half of the twelfth century. The walls, which still stand, are roughly 52 feet long and 26 feet broad; they are nearly 5 feet thick and rise as high as 13 feet. Apparently no mortar was used in their construction. The cathedral church of Gardar, dedicated to St. Nicholas—patron of seafarers—was considerably larger. It was just over 88 feet long and its greatest breadth was just over 52 feet. It was built of red sandstone, quarried from a neighbouring hillside. Traces of mortar have been discovered on the site. It would appear that the windows had glass panes. The church of St. Nicholas, which was orientated E.–W., was cruciform in plan. It had a narrow chancel and two chapels branching off it to north and south; the smaller of these chapels was used as a sacristy. From it a flagstoned walk led across the churchyard to the bishop's hall. In one of the graves at Gardar has been discovered the skeleton of a powerfully built, middle-aged man grasping a finely carved walrus-ivory crozier in his right hand and wearing a gold ring on the ring-finger of the same hand. This is believed to have been Jón Smyrill, Bishop of Gardar (1188–1209). Mention must also be made of another important discovery on the same site:

namely, the fragments of a great church bell, the memory of whose deep-toned chimes ringing out across the plain (it has been said) has survived to recent times in Eskimo lore.[9]

As time went on, the property of the see of Gardar gradually increased, farm after farm being added to the bishopric. By the fourteenth century the whole of the Einarsfjörd district belonged to the bishop, together with a number of large islands lying at the entrance to this fjord and the Eiríksfjörd, as well as extensive hunting-grounds on the east coast. Just as the Church in Norway had waxed steadily in wealth and power, so had the see of Gardar. It is significant that henceforward no more was heard of the chieftain at Brattahlid, who, in the earlier days of the Settlement, had always dominated the scene. In the later medieval era the great man in Greenland was undoubtedly the bishop, without equal or second. These developments could not but have had their effect on the Greenlanders. 'Richly endowed churches', Maitland observed long ago, 'mean a subjected peasantry.' It is at any rate within the bounds of possibility that this particular manifestation of *Popery and wooden shoes*—together with the very despotic governmental regulation of commerce and shipping—may have sapped the sturdy self-reliance of former generations and to that extent contributed materially to the colony's decline.

One of the most surprising features of Greenland agriculture is the immense number of cows that grazed in the once lush meadows, as evidenced by the number and dimensions of the byres which, in a ruined state, survive to this day. As has already been noted, a medium-size farm in the East Settlement would support as many as between ten and twenty head of cattle. The arrangements on the bishop's estate were on a proportionately greater scale. The fine aerial photographs which illustrate *Meddelelser om Grønland*, vol. 76 (1930), give some idea of the imposing dimensions of the episcopal farm with its massive stone byres capable of containing close on a hundred cows and its sheep-folds enclosing thousands of sheep. In the warm Greenland summers the flocks and herds would be driven up into the high mountain pastures. But in the last resort everything turned on the harvest. The survival of all this livestock depended upon the immense accumulations of hay and other fodder which could be laid up against the long cold winter.

The bishopric, then, was amply endowed, and the bishop lived in very good style. The episcopal residence was from time to time enlarged. The great hall at Gardar was the largest in the colony, measuring over 52 feet long and 23 feet broad. The hall was capable of holding several hundred people; here on great festivals the bishop sat in his high seat overlooking the crowded tables; here he would entertain the merchants from Norway after their long voyage across the ocean.

12

The Vinland voyages

THE *Grœnlendinga saga*, which furnishes such an interesting and lucid account of the first discovery of America by Bjarni Herjólfsson, continues with the narration of a whole series of successful ventures to that part of continental America which the Greenlanders called Vinland. The location of the region in question is as yet unresolved; and in all likelihood will never be known for certain. Nearly every historian who has ever attempted to solve the problem of its whereabouts has, in fact, come up with a different answer. The present writer does not propose to add to their number.

'The latest search for Vinland started in 1961', writes Björn Thorsteinsson (this was with reference to Helge Ingstad's recent investigations in Newfoundland); 'and it still remains an undiscovered country, as far as conclusive evidence goes.'[1] Nevertheless, in spite of all the uncertainty as to Vinland's geographical position, there is a strong and abiding fascination in these first vivid glimpses, of nearly a thousand years ago, of the American coastlands. The account which follows, like that of Bjarni Herjólfsson's voyage related above, is based on the *Grœnlendinga saga*.[2]

Fourteen years after Bjarni Herjólfsson has sighted the unknown coast away beyond Greenland, he went to Norway on a visit to Earl Eirík, from whom he received a cordial welcome. Bjarni told the earl about the voyage which had brought them within a short distance of the new islands on the other side of the ocean, which he and the others had seen but not visited. Because he had taken no steps to explore those new islands, Bjarni was strongly criticized by some for his lack of enterprise. However, he was given a place at the earl's court and the following summer sailed back to Herjólfsnes.

There was now much talk in Greenland about voyages of discovery. Leif, son of Eirík Raudi, made up his mind to seek out the land Bjarni had seen from the sea. He therefore bought Bjarni's ship from him and engaged a crew of thirty-four men. At the same time he appears to have learned all he could about the sailing directions for the voyage. Leif asked his father Eirík, despite his advancing years, to act as leader of the expedition; and Eirík, though not without misgivings, at last agreed to do so. When everything was in readiness the old chieftain rode down from Brattahlid to the shore where the ship was moored; but before he could get there his horse stumbled, with the result that Eirík took a bad fall and injured his foot. That settled the matter, so far as Eirík was concerned. He said he was evidently not destined to discover more lands than the one in which they were now living, and he would therefore part company with them. Eirík returned to Brattahlid, while Leif went on board the ship with his men.

They sailed down the fjord and out to sea on the course which they had learned from Bjarni. The first land they sighted was the one which Bjarni and his men had found last. They stood in to the land, where they anchored the ship, and pulled ashore in the boat. They saw that it was a bleak, flat, barren land of rock and ice, with no grass.

'At least', declared Leif, 'we have managed to get ashore, which is more than Bjarni did! I now mean to give the land a name and call it Helluland ("Flatstone Land").' After that they went on board again.

They stood out to sea again and sighted another coast, the second which Bjarni had found. Again they stood in to the land, anchored, and went ashore in the ship's boat. They saw that the country was low-lying and thickly wooded; and wherever they went there were broad stretches of white sand, which shelved gently to the sea.

'This land', declared Leif, 'shall be given an appropriate name, and shall be called Markland ("Wood Land").'

They hurried back to the ship and again stood out to sea with a north-east wind. They sailed for two days before sighting land again. They brought the ship past a cape, up a river, and into a lake, where they anchored and carried their sleeping-bags ashore, and built themselves huts. It was fine weather and the country appeared attractive; so they decided to winter there, and built good houses. After the house-building was finished, they explored the country around. They set to work gathering grapes and felling timber with which to load their ship. No frost came during the winter. There was fine salmon in the river, and the grass was so good that cattle would need no fodder in the winter months. Leif named the land after its produce, and called it Vinland, or Wineland.

In the following spring, with the ship fully loaded, they made ready for the homeward voyage and sailed away. They had a fair wind all the way until they came in sight of the mountains and glaciers of Greenland.

It was while they were approaching the coast of Greenland that Leif sighted some shipwrecked mariners stranded on a reef.

'Now what I mean to do', said Leif, 'is to work up to the reef if they should be in need of help and it is necessary for us to assist them; and if they are not peaceably disposed we have all the advantages and they have not.'

Directly they were within hailing distance of the reef, one of Leif's crew sang out to ask the castaways which of them was in command. 'His name is Thórir,' was the reply; 'and he is a Norwegian by descent.—What is *your* name?' Leif told him. 'Are you a son of Eirík Raudi of Brattahlid?' asked the other. Leif said he was. 'And now', he went on, 'I will take you all on board my ship, and as many of your belongings as the ship will hold.'

The men on the reef gladly accepted Leif's offer, and then they all sailed for Eiríksfjörd and arrived at Brattahlid, where they landed their cargo. Because of this timely rescue Leif was afterwards known as Leif the Lucky.

During the early years of the eleventh century the Greenlanders were well enough acquainted with navigational conditions on that part of the North American coast to be able to shape a course for Leif Eiríksson's landing-place

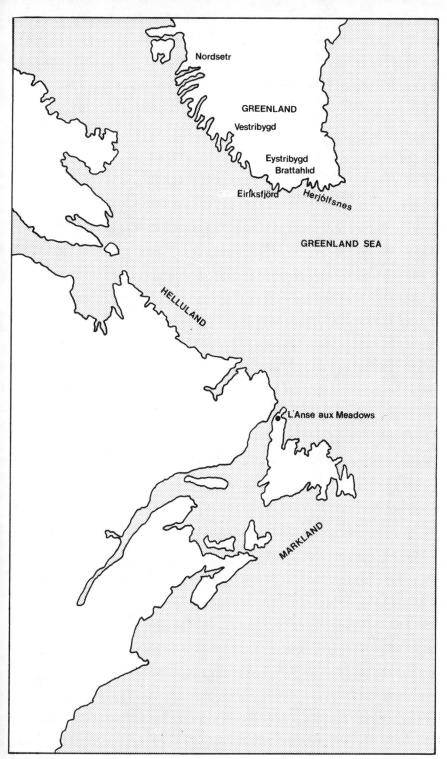

The Vinland voyages

without difficulty; and, as a sequel to Leif's enterprise, a whole series of voyages was presently made to 'Vinland the Good'.

After Leif's return to Brattahlid there was much talk in Greenland about his experiences in Vinland; and his brother Thorvald expressed his opinion that the country ought to have been explored more extensively. Thereupon Leif offered to let Thorvald have his ship for a voyage to new land; and Thorvald made ready for the expedition, taking with him, at Leif's suggestion, a crew of thirty men.

They put to sea and had a prosperous voyage, discovered Leif's lodgings in Vinland, and laid up their ship for the winter. Next summer they started to explore the country, returning to Leif's encampment in the autumn. The following summer Thorvald coasted along the more northerly part of the country, which they saw was densely wooded. 'This is a beautiful place', Thorvald observed, 'where I should like to make my home.' But coming suddenly on a party of natives (presumably Red Indians), the Greenlanders put them all to death with the exception of one man who managed to get away in his canoe. Following this affray, they were attacked one day by a much larger body of natives, and in the ensuing fight Thorvald was mortally wounded by an arrow. His crew later returned to Leif's encampment. The following spring they sailed back to Eiríksfjörd.

Thorvald's brother Thorsteinn now proposed to make the voyage to Vinland to recover his body. But fate was against him; the weather turned bad, and it is recorded that he spent the entire summer storm-tossed on the Western Ocean. Soon after his return to Greenland he died of pestilence.

The following summer a ship arrived in Eiríksfjörd from Norway. Her master was Thorfinn Karlsefni, an Icelander. Karlsefni was a man of substance and an experienced mariner. Once again there was talk of Vinland voyages, and Karlsefni (who had lately married Thorvald's widow, Gudrid) was persuaded to lead yet another expedition to that country. He secured a strong ship's company of sixty men as well as five women. They took with them all kinds of livestock; it being their intention to colonize the land if they could. They reached Vinland in safety, and set to work felling timber with which to load the ship, and also hunting and fishing. Next summer they were visited by a large number of natives, and trading began. Early in their second winter in Vinland, however, another party of natives came to the settlement, this time in even greater numbers. At first there was trading between the natives and the Greenlanders as before; but some time later a fight broke out, in the course of which the former were eventually worsted and put to flight. But there was no knowing when the natives would return, and, as word went round among the neighbouring tribes of the white intruders, in what might well be overwhelming numbers. That appears to have been the conclusion reached by Karlsefni; for when spring came he said he was not inclined to stay any longer in Vinland, and announced that he meant to return to Greenland. His men made ready for the voyage and put to sea. They arrived in Eiríksfjörd in safety with a valuable cargo of timber, hides, grapes, and other good things.

The last of the voyages made to Vinland during this period was that of Freydis Eiríksdottir—the illegitimate daughter of Eirík Raudi, and a bad lot. It was undertaken for much the same reason as the previous ventures, namely 'for the sake of profit and renown'. Immediately after their arrival in Vinland Freydis showed herself in her true light. The venture ended with the treacherous slaying of her two partners, Helgi and Finnbogi, and also of their unfortunate followers, both men and women. Freydis and her men returned to Greenland in the ship formerly owned by Helgi and Finnbogi. They sailed into the Eiríksfjörd in the early summer.

On their arrival Thorfinn Karlsefni was still there, with his ship ready for sea and waiting for a wind. It is said that a more richly laden ship never left Greenland than his.

So much for the *Grænlendinga saga*; the main features of which, if not all the details, would appear to be founded on fact. In the late Viking era Norsemen were certainly sailing to various parts of the American coast. The cargoes they were able to secure there evidently made the venture a profitable one. Markland and Vinland are not legendary islands, like 'St. Brendan's Isle'. Apart from the testimony of the *Grænlendinga saga*, there is clear and irrefutable evidence on the point. By a succession of chroniclers their existence is taken for granted.

The earliest evidence of all is to be found in Adam of Bremen's *Gesta Hammaburgensis*, in which explicit mention is made of Vinland. Adam received his information from Svein Estridsson, King of Denmark. 'Besides Iceland,' Adam related, 'there are many other islands in the great ocean, of which Greenland is not the smallest; it lies further away in the ocean. . . . Moreover, he [the King of Denmark] said that an island has been found by many in this ocean, which has been called Vinland, because there vines grow wild and bear good grapes. Moreover, that there is self-sown grain in abundance, we learned, not from mythological tales, but from reliable account of the Danes.'[3] The *Íslendingabók* of Ari Fródi, the earliest historian of Iceland, also contains a direct reference to Vinland.[4] Vinland is likewise mentioned in the *Landnámabók*,[5] and in a thirteenth-century Icelandic geography.[6] Lastly, in several of the Icelandic annals mention is made of both Vinland and Markland under the years 1121 and 1347 respectively.[7]

Two other possible sources of evidence have now to be considered.

First, there are the alleged Norse remains located, not only on the North American seaboard, but also far inland. The small cylindrical stone tower, known as the Newport Tower, in Rhode Island, has been mentioned in documents for at least three hundred years. Otherwise its age is unknown: nor is there any certain evidence that the building was of Norse origin; it is conjectured to have been some kind of watch-tower, or beacon, dating from about the middle of the seventeenth century. The Beardmore find, comprising a number of objects, of which the fragment of an iron sword is perhaps the most important, bears a close resemblance to various grave-finds of the Viking age from East Norway. The objects would appear to be genuine Norse antiquities. Whether or not they were taken to America in the early Middle Ages, however, is an altogether different matter. The balance of probability, indeed, is that they were brought there, not in the Viking era, but in comparatively recent times, by one Lieutenant John Block, a Norwegian. Best known of all the alleged Norse remains in North America is the Kensington Stone, which was discovered in 1898 in the vicinity of Kensington, Minnesota. For two generations at least it has been the centre of acute, and occasionally acrimonious, controversy. On one side of the stone, and also on one edge of it, are runes apparently cut with a chisel. The runes have been thus interpreted:

Eight Swedes and twenty-two Norwegians on an exploring journey from Vinland westward. We had our camp by two rocky islets one day's journey north of this stone. We were out fishing one day. When we came home we found ten men red with blood and dead. A V M [Ave Virgo Maria] save us from evil. We have ten men by the sea to look after our ships, fourteen days' journey from this island. Year 1392.[8]

Largely through the fervent championship of Hjalmar R. Holand,[9] this inscription has been accepted as genuine by a number of historians, archaeologists, geographers, and others; more especially on the western side of the Atlantic. On the other hand, it is not to be disputed that these runes on the Kensington Stone have been pronounced spurious by all the principal runologists of Scandinavia and Germany. In every significant detail—in runic forms, grammar, vocabulary, and even in the weathering of the runes—the inscription will not for a moment stand up in the face of expert criticism.[10] All the available evidence indicates, in fact, that the runes were carved at no earlier date than the final decade of the nineteenth century.

Notwithstanding the intense interest and excitement always evoked in the United States by any report of Norse artefacts, it could be confidently asserted, at any rate until quite recently, that no antiquities had ever been found in American which could be positively attributed to the presence of either Vikings or other Europeans in the medieval era.[11]

In the last few years, however, the investigations of Helge Ingstad at L'Anse aux Meadows in Newfoundland have resulted in discoveries apparently amounting to certain evidence. The excavations have brought to light the remains of eight houses (one of them a smithy), various boat-houses, cooking-pits, and a charcoal kiln. On one of the house-sites there was found a ring-headed bronze pin of a kind which was in common use in the Viking age for fastening the cloak or mantle. Another interesting find was a soap-stone lamp; also a quantity of bog-iron, the use of which was unknown among the aboriginals. Carbon analysis dates the house-site to the late tenth or the early eleventh century. There is general agreement among archaeologists that the remains at L'Anse aux Meadows are undoubtedly of Norse origin.

There may possibly be ancillary evidence of the link between North America and the Greenland colony in the shape of a quartzite arrow-head, believed to be of Indian origin, which was discovered in the churchyard at Sandnes in the West Settlement.[12]

The other possible source of evidence is cartographical.

It would appear that the same credulity which characterized the quest for Norse antiquities in America manifested itself again recently in the investigation of the Vinland question from the cartographical angle. This centred upon the discovery and acquisition by Yale University of the much publicized Vinland Map in 1966.

The naïvety with which the editorial panel appointed by Yale entered on their inquiries almost passes belief. One can say with reasonable confidence that, if at the outset they had enlisted the aid of Scandinavian and Icelandic authorities long experienced in this field, they would have been saved from a number of risky assumptions and quite unwarranted conclusions. Nor can the defence that they were precluded, under their terms of reference, from going outside their own small circle be well sustained in view of the fact that they are known to have approached, in confidence, at least one outside authority. As it was, they fell into error after major error arising out of their ignorance of Scandinavian history (which ignorance, incidently, was shared by the outside authority), of seamanship and navigation, and of various other specialist studies.

It is to be observed that some of the commentators who added their voices to the debate showed themselves scarcely less gullible than the several authors of *The Vinland Map and the Tartar Relation.*

In the absence of supporting evidence, the delineation of the Greenland coastline, both east and west, is accurate to such a degree as to arouse immediate suspicion. The question at once arises, when and how did the cartographer get the necessary information? According to Professor Gwyn Jones, speaking at the Vinland Map Conference in 1971, the delineation of Greenland appears to owe everything, 'to first-hand knowledge' and in fact 'represented accumulated Norse experience'.[13] The maritime evidence by no means sustains this view of the matter. There are no reasonable grounds for supposing that the circumnavigation of Greenland was ever achieved at all in the medieval era, though it is certainly implied in the Vinland Map.

Another problem arises with regard to Vinland itself. Whence came the knowledge of its position and general outline? Such knowledge, as has been said, did exist in the early part of the eleventh century; the sailing directions were known and followed for several years; but in the course of the next century or so that knowledge was lost, and, so far as is known, it was never recovered. 'Markland', Thorsteinsson sums up, 'was one of the certain facts of geography which the Greenlanders, and thus the Norwegians and Icelanders, possessed on the Middle Ages: Vinland was not.'[14]

The erroneous conclusions into which the members of the panel were apt to be betrayed through their ignorance of a language indispensable for research in this field—namely, Old Icelandic—is illustrated, in the well-known entry in the Icelandic annals of Eirík Upsi's Vinland expedition of 1121, by the writer's misinterpretation of the verb *leita*. *Leita* in this context, assuredly, does not mean 'lay a course for' the land in question, but 'to go in search of.' His failure to grasp the true meaning of *leita* in this passage led inevitably to serious and cumulative errors in the chain of reasoning, leaving the writer at last caught up in a flight of pure fantasy with speculations concerning the existence of a Norse colony in Vinland and their possible contacts with compatriots in Greenland.[15]

It remains to say that all this arguing and theorizing went on without any real attempt to establish the provenance of the document which had been acclaimed by one of the editorial panel as 'the most tremendous historical discovery of the twentieth century'. The whole long-drawn-out discussion resolved itself into a series of academic exercises, making little or no contribution to history. Suspicions of forgery, so far from being dispelled, progressively increased and intensified; and in face of the very formidable body of criticism which arose, first, in Scandinavia and Iceland, and, later, in this country and elsewhere, from the cartographical, philological, and historical angles, 'the most tremendous historical discovery of the twentieth century' finally toppled down like a house of cards.

There is every reason to believe that in the later Middle Ages the Greenlanders ventured occasionally to Markland; though probably not to Vinland. As late as 1347, the year after Crecy, the Icelandic annals record: 'A craft came from Greenland which was smaller in size than a small Icelandman. It came into the entrance of Straumfjörd. It had no anchor. There were seventeen men on board who had been on a voyage to Markland and later had been driven by gales to this land.'[16] As Reman rightly observes, 'the Markland voyage could not have been made unless oral tradition in Greenland had preserved the memory of the course thither'.[17] So great was the Greenlanders' need of sound timber that, even when such craft as remained in the colony were really neither large nor seaworthy enough for the venture, they would make the voyage to Markland; the sailing directions for which were, apparently, still known to them.

Nor is there anything like the same element of doubt concerning the approximate location of Markland, as there certainly is in the case of Vinland. 'Southward from Greenland is Helluland,' records a medieval Icelandic geographical treatise, 'and after that Markland. Thence it is not far to Vinland the Good.'[18] Nevertheless, despite the last confident assertion, there are really no grounds for supposing that a century or so after the series of voyages made by Leif Eiríksson and his contemporaries, the course for Vinland was still known to the Greenlanders. It is significant of much that, in 1121, the Icelandic annals record that 'Eirík, the Greenlanders' bishop' went in search of Vinland ('Eiríkr byskup af Grœnlandi for at leita Vinlands').[19] That is, the bishop went in search of a land that was known to exist; though the knowledge of its approximate situation, and the way there, had apparently been lost. Moreover, with this laconic entry in the Icelandic annals Eirík disappears from history. It is unknown whether he reached his destination or whether he ever returned. Three years afterwards Arnald, as we know, was appointed bishop of Gardar.

There is no record of any *planned* voyage from Iceland to America or vice versa. Whenever such a visit was made, it was through stress of weather. The organized ventures to Markland and Vinland were always made from Greenland. This was destined to have decisive consequences. The truth is, the resources of Iceland, in ships and men, had sufficed for the colonization of Greenland: but the resources of Greenland were utterly inadequate for such an enterprise as the settlement of Vinland. Judging from the experience of Karlsefni and the others, a substantial body of well-armed warriors would have been required to hold the hostile natives in check. The various expeditions which set out from Greenland were not nearly strong enough to establish a permanent settlement in such conditions. Another and most serious handicap was the lack of shipping. It is scarcely to be wondered at that the American mainland was seldom visited if the Greenlanders possessed no better craft for the passage than the small vessels referred to in the Icelandic annals in the years 1189 and 1347.[20]

This last, and, as it proved to be, abortive manifestation of what Brøgger has described as 'the drive to the West', *vestmannamotivet*, had no such decisive influence upon the course of world history as the voyages of Columbus and the others in the age of the great discoveries. The failure of the Norsemen to establish a permanent settlement in North America is well summed up by Gathorne-Hardy. 'Karlsefni and his contemporaries', declares this authority,

were—as discoverers—born out of due time. With the general interest which was felt in exploration in the fifteenth and following centuries, with kings to back them and states to develop their discoveries, above all, with an armament immeasurably superior to that of the natives, such as the later explorers possessed, these simple Norse seamen might have attained a far wider fame, or even have affected the course of history. As it was their deeds were unimportant, and soon almost if not quite forgotten.[21]

13

The Norse traffic to Iceland

FROM THE TIME of the Settlement a flourishing trade had sprung up between Norway and Iceland. In these early days the principal imports of Iceland, according to the sagas, were timber and meal; the chief export was *vaðmál*, or wadmal, the coarse woollen cloth of the country. Later the Icelanders began also to import wine, beer, malt, wax, incense, honey, woollen cloth, linen, tar, tin, copper, iron, and iron goods; and they exported hides, sulphur, falcons, train-oil, and, above all, *skreið*, or stockfish. To and from Iceland came shipping from all parts of the Scandinavian world—Norway, Denmark, Ireland, the Sudreyjar, the Orkneys and the Shetlands, the Faeroes, and Greenland.

During the emigration period, and for several generations afterwards, many of the Icelanders were bold and experienced seamen. It was Icelanders who colonized Greenland and later ventured to various parts of continental America. It was in all probability an Iceland crew which, in 1199, discovered Svalbard (believed to have been some part of East Greenland). Down to the early half of the thirteenth century the Icelanders possessed their own trading-fleet. In 1131 one of the three *kaupskip* visiting Greenland was an Icelandman. It was merchants from Iceland who, about 1200, set up the beacon on Sanday Island in the Sudreyjar. In the early Middle Ages they frequently sailed to Ireland and England and, as late as 1225, we hear of vessels from Iceland at Yarmouth (though it is possible that these last were actually Norwegian). Voyages to the Sudreyjar and the Orkneys are often mentioned; and in 1198 an Icelandic craft visited Rouen. For the most part, however, they sailed to Norway, and especially to Trondheim. For two centuries and more after its foundation at the close of the tenth century, this town was the centre of the Iceland trade. In the time of Magnus Berfœtt (1263–78) there were no less than 300 Icelanders living in Trondheim.

After about 1250 the traffic to Iceland, as well as that to the Shetlands, Faeroes, and Greenland, was almost entirely in the hands of the Norwegians. Norwegian vessels and Norwegian crews are very frequently mentioned in the thirteenth-century sagas. At a much later period, however, the Icelanders still owned a few ships. These belonged to certain of the more important personages, such as a bishop or a lawman.[1] Thus Vilchin, Bishop of Skálaholt, towards the end of the fourteenth century, owned his own craft. 'He had also a *búza* built in Norway, which was called the Bishop's *búza*. She was always a fast sailer, as was to be expected.'[2] There were also a few vessels built in Iceland out of timbers shipped from Norway, or from other ships broken up. For the rest the Icelanders came to rely on Norwegian shipping.

There was no internal trade of any importance in these early centuries. With very few exceptions all import and export trade was done with countries overseas. It is to be added that this traffic was largely—though not entirely—a trade in luxuries and other commodities which were costly in comparison with their bulk. The fact is, until the arrival in Iceland of substantial English and Hanse merchantmen in the fifteenth century, the vessels which made the Iceland voyage were never very large.

Timber was one of the oldest imports to Iceland. From the time of the Settlement timber for house-building often formed part of the cargoes which were brought out to Iceland from Norway.[3] During the thirteenth century the Icelanders sometimes fetched it for themselves. It would appear that the timber trade was fairly well organized at a comparatively early date, seeing that in the sagas a distinction is made between ordinary timber (*víðr*) and timber of prime quality (*kjör-víðr*).[4] Apart from drift-timber[5] there were extensive birch-woods in various parts of the island at the time of the Settlement. But heavy timber for church- and house-building had to be fetched from across the sea. The earliest voyages were made between Iceland and the western coast of Norway; but from the eleventh century the Icelanders also sailed to Vík. In later centuries much of the timber exported to Iceland came from Vík. Though the total quantity of meal exported to Iceland in any one year can never have been very great it was an important factor in the island's economy. It is recorded that, in a time of dearth, Harald Hardrádi dispatched four ships laden with flour to Iceland.[6] The Icelanders had need of these cargoes, few though they were. For centuries the merchantmen sailing from Iceland to Norway were laden with bales of *vaðmál* woven from the wool of Iceland's abundant flocks.[7] Among other early exports of Iceland were hides. In the *Njáls saga* and other sagas mention is made of vessels from Iceland 'laden with skins and furs'. Falcons, too, are sometimes mentioned as coming from Iceland. According to Cleasby and Vigfusson's *Icelandic–English Dictionary* the foreign word, *fálki*, derived from the Latin, came into use as a trade term and only occurs in the thirteenth century. The white falcon (*F. islandicus*) was much sought after in the Middle Ages.[8] Among the *xxij girofalcones* presented to Henry III by Hákon Hákonarson in 1225 were three white falcons which must have come from either Greenland or Iceland. In 1262 the Sultan of Tunis received a present of Icelandic falcons.[9] In the era of Hákon Hákonarson the gerfalcon industry became a royal monopoly and the eyries in Iceland were guarded by royal officials.[10] Of *skreið*, which was to be the future mainstay of Iceland trade, we hear but little until the later Middle Ages. In the early years of the fourteenth century there is only a single reference in the English customs accounts to 'fish from Iceland'. In 1340, however, there is a significant passage concerning the increasing demand for stockfish in connection with the quarrel between the Iceland merchants of Bergen and the Archbishop of Nidarós over tithes. The former, who had been accustomed in the past to pay tithes on the bales of *vaðmál* imported into Norway, had been unwilling to pay them likewise on *skreið* and train-oil, which were now regarded as the principal exports of Iceland. In the latter half of the fourteenth century immense quantities of stockfish were exported to Norway. So much so, in fact, that the Icelanders took alarm at the huge cargoes of *skreið* which were leaving their shores, and protested to the government of Norway–Denmark that they were left without sufficient supplies for their own consumption. A large proportion of this *skreið* was re-exported to Germany and England.[11]

Notwithstanding their insular position the Icelanders were very well informed about northern affairs,[12] and travelled extensively in Europe. Two of their bishops, Thorlák and Pál, were educated in England. There were not a few merchants who lived partly in Iceland, and partly in Norway. Icelanders of good family were accustomed to send their sons abroad to complete their education. The bishops of Skálaholt and Hólar journeyed freely to and fro between Iceland and Norway. There is nothing, in fact, to suggest that the ocean passage was ever regarded as a very hazardous undertaking. In the Icelandic annals many references occur to pilgrimages made by both clerics and laymen. Rome and Jerusalem are frequently mentioned.[13] The famous shrine of St. James of Compostela, however, was apparently but rarely visited. This pilgrim traffic lasted as long as the ocean voyages of the Norwegians. As late as the year 1406 it is stated that the son of an Icelandic chieftain, Björn Einarsson, crossed the ocean with his wife Solveig and went on pilgrimage, first, to Rome, and, later, took passage from Venice for the Holy Land. After their return from Jerusalem Björn set out for the shrine of Compostela, while his wife went back to Norway.[14]

The passage to Iceland was usually made in the late spring or early summer, 'when the sea began to go down', and the winds might be expected to be easterly and fair for the voyage. 'There came two ships to Grunnasundnes before the *Althing*.' 'There came a Faeroese craft from Shetland to Iceland, to Önundarfjörd in Vestfjörd, before St. Vitus's day.' 'That summer thirty-five ships sailed to Iceland.' 'Thórd put to sea in early summer and arrived in Gufuárós, for the *Althing*.' 'When it was spring, men began to fit out their craft for trading ventures to various lands.' 'Thórgeir fitted out his ship in early summer to sail to Iceland.'[15] The Norway–Iceland passage was sometimes made in the autumn as well. 'He came out in the autumn to Eyri.' 'There came out no more ships from Norway to Iceland that autumn.' 'There came out the two Bishops, Thórarinn and Jón, and they came ashore before Michaelmas.' 'There arrived the new Governor, Andrew Sveinsson, . . . in Straumfjörd before the Winter Nights [24–6 October].'[16] The eastward passage to Norway was frequently made in the late summer or autumn. 'The comet was sighted at Michaelmas by sea-traders, who were on the high sea.' 'Eight ships came to Norway, in the autumn, after Michaelmas.' 'They were late in getting a fair wind and arrived off Hálogaland after the Winter Nights.' 'He made the passage to Norway that autumn and came to Norway a little before the Winter Nights. . . . He went overseas in the autumn, and passed the third winter in Norway.' 'They made up their minds to sail from Gáseyr with Hallstein the Humpback and stood out to sea on Michaelmas Eve.'[17]

It would appear from the thirteenth-century sagas that the principal sea-routes at this time were those from Trondheim to Eyjafjörd, from Cape Stad to Horn, from Bergen to Eyri and the Vestmannaejar, and from Bergen to the havens on the west coast of Iceland. Eyrar, Hvita, and Gáseyr were, perhaps, the havens most frequented during the commonwealth period (885–1263). Vessels bound for the south, west, or north of Iceland apparently made the land at Horn or some other point on the east coast before proceeding to their destination.[18] The usual route to the havens on the west coast was along the south coast and round Reykjanes; on this route a craft would have the advantage of the current, which sets in a clockwise direction round the coasts of Iceland.[19] Even in the time of the Settlement, however, the north-about route was sometimes used.[20] In the later Middle Ages, when Trondheim was gradually superseded by Bergen as the chief centre of the

Iceland trade, many ships bound from Iceland to Norway no longer sailed to Trondheim, but to Bergen. Thus in the *Laurentius saga* it is related that, 'in spring, after Easter, Sir Egil came before the archbishop, and said that time was flying, and that he had a long journey before him, first south to Bergen and thence to Iceland; there being no passage to Iceland then from Trondheim. . . . South to Bergen went Sir Egil, and lit on a ship trading to Iceland. They had a fair voyage, and reached Eyri safe and sound before St. Laurence's Day.'[21] In the fourteenth century Hvalfjörd, Skutilsfjörd, Dýrafjörd, Kolbein, and Gáseyr in the north, and Gautavík, Eyrar, and the Vestmannaeyjar in the south, were the Iceland havens generally used. Little is known for certain about the course shaped between the two countries. As a rule vessels homeward bound appear to have sighted no land at all until they came in sight of the mountains of Norway. Occasionally, however, they might sight the Faeroes or one or other of the Northern Isles, such as Fair Isle, during the passage.

The statistics of the traffic between Iceland and other lands, as recorded in the Icelandic annals, should always be viewed with caution. Certainly the entry for any particular year is no safe index to the comparative frequency of communications between the two countries. As a rule it is better to take the average for a decade or so. There can be no doubt that, in some years, a substantial proportion of the sailings to and from Iceland never appeared in the annals. On the other hand, when they definitely mention a given number of sailings between Norway and Iceland[22]—or state that no vessel at all came out that season—the evidence may be accepted as fully trustworthy. Sometimes the fact that no ship arrived from Norway throughout the year is of no particular significance. No ship arrived, for instance, in the years 1187, 1219, and 1326. Yet the commerce between the two countries, as we know from a variety of sources, prospered throughout this period. But the record of not a single sailing in the year 1350 *is* significant, because it reflects the decline of Norwegian commerce and shipping accelerated by the Black Death. From time to time references occur to a ban, total or partial, on the Iceland trade.[23]

There are occasional references in the Icelandic annals and sagas to the number of persons aboard any one craft. For the most part no comprehensive list of passengers is given; but only the names of the more important personages who made the voyage—for example, the bishops and other clergy, the governor, and various magnates. It is not very often that the master's name appears on the list. The *Byskupa sögur* and the *Sturlunga saga* frequently give details which are not to be found in the Icelandic annals. From all these sources it is apparent that there was a constant coming and going of ecclesiastics, chieftains, officials, and merchants between Iceland and Norway.

'There sailed from Norway Lord Álf from Króki, Lord Sturla Jónsson, Lord Erland, and many other Icelanders.' 'A ship arrived from Norway in Kolbeins estuary on the north coast. The master's name was Ívar.' 'Bishop Orm sailed for Norway from Hvalfjörd. There sailed also the priests Einar Haflidason and Runólf. They both travelled at the expense of Lord Bishop John.' 'There sailed from Norway Bishop Michael, Brother Thorstein, and many priests. There sailed also Andrew, the governor, Björn Einarsson, Sigurd of the White Locks, and Beaky John.' 'In the following spring Thórgeir . . . joined a ship in the Eyjafjörd belonging to Ögmund Amberhead, who was the father of Helgi, who was later Bishop of Greenland. Thórgeir was accompanied by Thórólf the priest, son of

Snorri, Thorstein, Thorkel, the son of Eirík, and many other Icelanders.' 'Besides Gudmund, there were fifteen Icelanders sailing in this ship; these were Hrafn, the son of Sveinbjörn, Tomas, the son of Thórarinn, Ívar, the son of Jón, Grím the monk, son of Hjalti, Erlend the priest, Berg, the son of Gunsteinn, Eyjólf, the son of Snorri, Thorstein, the son of Kambi, Gudmund the priest, son of Thormód, Brand, the son of Dalk, Petr, the son of Bárd, and his brother Snorri, Thórd, the son of Vermund, Höskuld, the son of Ari, and Kollsvein, the son of Björn.' 'And when Gizur had spent three winters in Norway, he got permission from King Hákon to sail for Iceland. All these went out in one ship: Bishop Henry and many Icelanders; Thorgils skarði Bödvarsson, Finnbjörn Helgason, Árnor Eiríksson, and many other Icelanders. That ship sailed into Eyjafjörd at Gáseyr. The master was Eystein the White.'[24]

A considerable number of ships are mentioned by name in the Icelandic annals and sagas.[25] There was the *Stangarfoli*, which in 1189 was driven off her course and wrecked on the Greenland coast.[26] There was the *Krossbúzuna*, to which references occur in the Icelandic annals in the years 1312, 1315, 1331, and 1334.[27] There was the *Langabúza*, whose arrival at Gáseyr is chronicled in 1335, and whose wreck is recorded in 1342.[28] There was the *Bauta hluti* (master, Sigurd Kolbeinsson), which was fitted out for the Greenland trade in 1336 and made a passage from Norway to Austfjörd in 1388.[29] There was the *Sunnifusúðin*, which came to Gáseyr in 1370, which lay in the Siglufjörd in 1373, and which was lost with all hands in 1388.[30] Although such references are fairly numerous, it would be rash indeed to attempt to deduce therefrom the average life of a *hafskip*. It would appear, however, that a *hafskip* might very well stay afloat for something like thirty years.[31]

Ships, and shares in a ship, were freely bought and sold. It occasionally happened that a vessel would be owned half by Norwegians and half by Icelanders; or again, half by Icelanders and half by men of the Sudreyjar. Many such instances are recorded in the sagas.[32]

The *hafskip* was essentially a sailing-vessel. Often we hear of crews waiting, sometimes for weeks on end, for a fair wind.[33] Their dependence upon the seasonal easterlies and westerlies can be gauged from the recurrence in the sagas of the ominous phrase *tók af byri*, 'the fair wind failed'. Along the coasts of Norway and Iceland there are a number of promontories with the significant name of Stad which give testimony of the numerous occasions on which a vessel would be obliged to lie and wait for a change of wind.[34] Distances, too, were reckoned in terms of 'day's sailing'. (See p. 109.) According to Adam of Bremen, the passage from Denmark to England took about three days' sailing. From Bergen to the Shetland Islands would take about two days. From the Faeroes to Iceland would take about a week. From Norway to Iceland would take about three weeks. The Greenland passage, of course, was a far more formidable matter. At best it might take several weeks; at worst it might run to as many months.

There were in all three types of *hafskip*. The *knörr* was the usual ocean-going ship of the Norsemen in the early Middle Ages, and is commemorated to this day in numerous place-names: for example, in Norway, Knarberg, Knarboe, Knardal, Knarested, Knarfjeldet, Knarlag, Knarreviken, Knarrum, Knartarud, Knarvaag, Knarvestol, Knarvik, Knurdalen, Knurlien, Knurren, Knurrested, Knurøen, Narraviken, Narvigen, Narvik, Narviken: in Iceland, Knararbrakkutangi,

Knararos, Knarrareyrr, Knarrarnes, Knarrarsund, Knarrarvad, Knorringfjellet: and in the Orkneys, Knarstoun. The Icelandic sagas mention certain individuals associated with the *knörr*. For instance, in the reign of Ólaf the Saint there was a shipwright called Thorstein Knarrasmid, who was a renowned builder of *knerrir*;[35] and in the days of Hákon Hákonarson there was a shipowner named Knarrar-Leif, who was entrusted by that monarch with an important mission to Greenland.[36] It was the *knörr* which in the Viking era and early Middle Ages was used for the navigation to the Faeroe Islands, Iceland, and Greenland.[37] The Icelandic sagas, though written at a time when the *búza* was coming into common use in the North, rightly allude to the *knörr* as the normal ocean-going merchantmen of the Viking era. On most of the northern sea-routes, in the thirteenth century, the *knörr* was supplanted by the *búza* and the cog.[38] But on the Greenland passage no other type of *hafskip* is recorded down to the end of the Middle Ages.

There can be no doubt that some difference existed between the two types, for the terms *knörr* and *búza* are never used of one and the same craft. It is very likely, as Brøgger has suggested, that the difference lay in the ship's stem. But there is no certainty on this point. The *búza* was a longship originally, and was used as a warship. Harald Hardrádi's *Long Snake* was a *búza*.[39] But many references occur in the Icelandic annals, the *Byskupa sögur*, the *Sturlunga saga*, and other Icelandic sagas, to the *búza* used as a merchant vessel.[40] Most of the Norwegian vessels which came to England in the early years of the fourteenth century were *búzur*. The third type of ocean-going craft was the *byrðingr*, which was short and broad in the beam, and somewhat smaller than either the *knörr* or *búza*, and with a smaller crew, comprising only about a dozen men. The *byrðingr* is often mentioned, particularly on the Norwegian coast, as a freight-ship or coaster. It was largely employed in the carriage of stockfish from the Lofoten Isles to Bergen. The *byrðingr* is also mentioned as venturing occasionally to the Faeroes and to Iceland;[41] but never to Greenland.

In none of the craft of the Viking era which have survived has any trace been found of a *vindáss*, or windlass. A statement which was made in Falk and Shetelig's *Scandinavian Archaeology* to the effect that in the Gokstad vessel the supports for the windlass were 'found in front of the mast' is altogether mistaken; nor does the error reappear in the later *Vikingeskipene*.[42] The absence of a windlass may be due, as the late Captain Sölver has suggested, to the fact that on board the Gokstad vessel and other craft of comparable size it was not really necessary: the anchor could easily be hoisted up by strength of hand.[43]

In the sagas reference is made to two different kinds of anchors: the *stjóri*,[44] which was evidently a killick, and the *akkeri*, which was a real anchor made of wrought iron and fitted with a wooden stock. The *akkeri* found in the Oseberg ship bears a marked resemblance to the iron anchor of the Mediterranean.[45] The name as well as the form, indeed, betrays its foreign origin. In Old Norse it is always *akkeri*; in Old Swedish, *akkæri*; in Old Danish, *akkæræ*; in Anglo-Saxon, *ancor*; in High German, *ancher*. The *akkeri*, like the sail (*segl*), was an importation from the south. How and when the iron anchor was introduced into Scandinavia, however, is unknown.[46]

During the excavation of the Viking ship at Ladby in Denmark in 1937 the discovery was made, in the forepart of the vessel, of an iron anchor complete with anchor-chain and anchor-rope, which casts a good deal of light on the technical skill and seamanship of the age. Like the Oseberg anchor the Ladby anchor has both an anchor-ring and a buoy-ring. In certain other respects as well it closely resembles the Oseberg anchor, though it is approximately three times as large. A distinctive feature of the Ladby anchor, however, may be seen in the 'small, strong flukes at the ends of the arms'. Sölver in an interesting paper has testified to the fine craftsmanship of this relic of the Viking era. 'Even by modern standards the anchor of the Ladby ship is a very good piece of ship's equipment and at the same time an unusually fine piece of handiwork. . . . Forging and welding seem to be perfect in every detail; the various parts are handsomely and accurately proportioned and symmetrically planned, and the uniform gauge of the various members—not obtained by filing, be it noted—is a feature of the whole anchor.' On the crown of the anchor is a ring (*hnakkmiði*) like that of the Oseberg ship for a buoy rope (*hnakkiband*), to which the anchor-buoy was secured. What makes the Ladby anchor a unique discovery, however, is the chain forerunner. This, as Sölver has declared, is the product alike of considerable technical skill and fine seamanship.

A chain forerunner on the anchor rope such as we now see for the first time in the history of anchors, was an extremely practical device. The weight of the chain keeps the anchor from lifting from the ground, the pull on the mooring is much more elastic than when the whole anchor cable is made of rope. But the greatest advantage of the chain forerunner is that the mooring is not frayed on rocks and stones on the sea bottom after the ship has lain at anchor in a current way. This chain forerunner must have been an uncommonly effective safety device when the ship was anchored, one example among many to show how practical and ingenious the seafarers of those days were with their gear.[47]

It is possible to form some conception of the general appearance and structure of these *hafskip*, or ocean-going merchantmen, from certain passages in the Icelandic sagas; notably in the *Egils saga* and the *Vatnsdœla saga*. It is related in the *Egils saga* that the ship belonging to Thórólf was well found in every respect; it was beautifully painted down to the water-line; the sails were striped with red and blue; and all the tackle was as well fashioned as the ship itself. One of the fastest sailers on record in the Viking age was the *Stígandi*, mentioned in the *Vatnsdœla saga*, which Harald Hárfagri presented to Ingimund.[48] The *Stígandi*, so-called because she 'strode over the waves', appears to have been a small, well-built vessel with very fine lines and of the traditional light flexible design.

The knowledge derived from the literary sources can be usefully supplemented by the archaeological evidence recently furnished by the discovery of the Skuldelev ships.[49] These wrecks, which were sunk in the fairway near Roskilde, Denmark, apparently to serve as blockships, included more than one merchant vessel; but 'Skuldelev I' is the largest and most important of them. The vessel has been shown to date back to the late Viking age. It was a strongly built craft of from forty to fifty tons, nearly 66 feet in length, with a deep, roomy hull and a high freeboard. The keelson, keel, and frames were of oak, and it was planked with pine. The planks were clinchered together with iron nails, and the planking was fastened to the frames with trenails.

It was a broad deep vessel with a marked division between the bottom of the ship terminating at the fifth strake, and the broad rounded bilge followed by the almost vertical upper strakes. . . . Heavy cross-beams attached to strong horizontal knees above the forward half-deck, and also presumably at the mast and above the after deck, acted as a considerable reinforcement for the ample hull above water. . . . In short, the primary concern of those who built the ship, with its strikingly roomy and sturdy hull, was to achieve the most solid construction possible.

A point of particular interest is that the system of flexible lashings characteristic of these larger Viking craft was abandoned, in the case of 'Skuldelev I' (as in that of the earlier Äskekärr vessel), in favour of a method of rigid fastenings. Even so, the method of fastening the strakes to the ribs with trenails was used in such a way as to allow the structure some measure of flexibility.

Clear proof of an advanced sailing technique was found forward . . . where a strong cleat with three indentations facing obliquely forward was fitted. This cleat served to secure a *beiti-áss*, the tacking spar, in three different positions when the wind was abeam to port or across the port bow.[50]

Skuldelev I, therefore, was essentially a sailing vessel. There were only a few oars fore and aft, and these apparently were little used, since they showed no signs of wear.

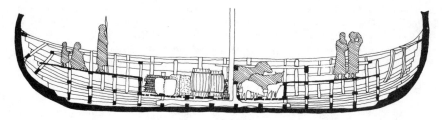

'Skuldelev I' (reconstruction) (Viking Ship Museum Roskilde)

From the numerous voyages recorded in the Icelandic annals, the *Sturlubók*, the *Byskupa sögur*, and other sagas certain conclusions may be drawn. In most years a large proportion of the ships engaged in the Iceland trade must have reached their destination without difficulty. Thus in the *Njáls saga* a crew set out from the Sudreyjar for Iceland: 'Now they put to sea and made a fast passage; and they got a fair wind, and made the land at Eyrar.'[51] In the *Laurentius saga* a crew sailed from Trondheim: 'They put to sea, they had a prosperous voyage, and landed safe and sound at Gáseyr.'[52] Elsewhere in the *Byskupa sögur* it is recorded that, 'When it was summer Thorlák . . . put to sea with his crew, and they had a prosperous voyage, and arrived in the port for which they were bound.'[53] The time usually taken on the run between Norway and Iceland was about three weeks.[54] But sometimes luck was against them and they made a long passage. 'Lawrence went to Gáseyr, and so on board. The traders put to sea nigh on St. Bartholomew's Day. They had a poor wind and were long at sea. At first they went north towards Naumadale, and then they had a good south wind and got into Trondheim in autumn nigh on Michaelmas Day.'[55] There are fairly frequent references to long passages,[56] and of vessels driven back to port by strong head-winds: in the fourteenth century alone many such instances are recorded.[57] Again, in a good many cases disaster in one form or another overtook the merchantmen plying between Norway and Iceland. Sometimes a vessel sailed out across the ocean and was never heard of again.[58] Sometimes the loss of ships would be made known through the reports of survivors or eyewitnesses.[59] The losses were as a rule recorded when they occurred in the vicinity of Iceland. Elsewhere it was another matter.[60] There were notably disastrous years like 1118, when the losses were heavy, and 1183, which became known as the Disastrous Summer, when 'five hundred men in ocean-going vessels were lost'.[61] Another bad year was 1343.[62] On the other hand, in some years the proportion of losses was comparatively low.[63]

In the *Gudmundar saga* there is a detailed account of a shipwreck off the north-west coast of Iceland. There the coastal and foreign-bound shipping was obliged to skirt the barren and inhospitable shores of the Strandir, fringed for about twenty miles by rocks, reefs, and shoals extending for miles offshore, where the depths are in many places so shallow that line upon line of breakers are formed. The ship which was carrying the future Bishop Gudmund had got in among the heavy overfalls off the Strandir.

The wind took them north of Gnupar off Melrakkaslata, but then came a head-wind; they let the ship drift and thus they were tossed about by it for a long time and driven westward to Hornstrandir. One Saturday evening, as they were sitting at supper, a man called Asmund, a Norwegian, threw open a corner of the tent and cried: 'Hi! hi! Off with the awnings! Get up, men, and be quick about it! Breakers ahead! Shove aside the tables and never mind your suppers!' They all sprang up and threw off the awnings. . . .

But now they were almost on the breakers and they wrangled greatly about what to do, and everybody wanted his own way. Some wanted to hoist sail and rushed to do so. . . . Just as they had lifted the sail less than six feet from the freight, a heavy sea dashed against it, fore and aft, and broke over it. There was someone clinging to every rope. Ingimund seized a boat-hook and tried to pull down the sail, while his nephew Gudmund had his place in the ship's boat, and had to try to disentangle it as he stood between the boat and the sail. But just at that moment there came another heavy sea, so great that it went over the whole ship. It swept away the vane of the mast and both the shelter-boards, and overboard went the sail and every part of the cargo that was not fastened down, except the men. The ship was much damaged and so was the boat. They got clear of the breakers when a third sea struck them, and this was not so heavy as the others. Now the men rushed to bail out the ship, both fore and aft, and the sail was hoisted. Then they saw land and began discussing where they were. . . . Now the ship was driven north of Reykjafell, but then no further, and they reefed sail and cast anchor. After a time the anchors held and they lay there for the night. The next morning men got ashore on planks of ship's timber. . . . Then the ship heeled over away from the shore, and everything in it was carried out to sea; the ship itself broke up and only a small part of the cargo drifted ashore.[64]

Since the mariner in the medieval era and long after had no means of determining his longitude except by rough and ready reckoning, he often struck the coast, when running his distance, before he realized the danger. It is related in the *Laurentius saga* that Lawrence and his party set off from Gáseyr in Eyjafjörd on 5 September 1323.

They had a good wind north off Langanes, and also eastwards over the sea; a good strong breeze and keen weather. They were only a little time and touched Hálogaland, in the north, opposite Brunney. Where they were going was an inshore reef, but they saw where they had got to and launched a boat. But the trader dashed up on the skerry quicker than they thought; and the ship split from under, and straightway the keel was down, and all the freight sank. The lord-bishop-elect, all the women, and all the less fighting folk went into the boat; some got to land with the mast-tree. All the people reached land, save that one woman died on the spot, named Thórdís, and known as 'Blossom-Cheek'. The whole fund of money mostly perished instantly on the spot; yet a good deal was saved, for they dragged up the rolls and bales of wadmal and the casks of train-oil; and of dried fish there was none on board.[65]

A quite common cause of shipwreck must have been overloading. The classic example of this, perhaps, was the fate of Flósi and his craft in the *Njáls saga*. Flósi set out from Norway, despite the remonstrances of his friends, late in the season and in a ship which was overloaded with timber; neither he nor any of his men were ever seen again.[66] A similar case is recorded in the *Thorláks saga*. When Bishop Thorlák was ready to sail for Iceland, he found that the crew had loaded the ship with deck timber until she could hardly float.[67]

It is worth noticing that the majority of shipwrecks occurred in coastal waters rather than on the high sea. The Norwegian skerries, the roosts of the Shetlands, the rocks and currents off the Faeroes, the fog-bound eastern coast of Iceland, claimed many a victim. 'In the autumn the *Krafsinn* was wrecked on the coast of Hálogaland.'[68] 'They drove to the land and sailed one night among the Shetlands. There was a heavy tide, they fell among the breakers and the ship broke up.'[69] 'The *Maríusúðin* was lost near the Shetlands with her crew and all her lading.'[70] 'The *búza* . . . was wrecked off the Faeroes with the loss of fifty men.'[71] 'Gudmund Ormsson was lost off the Faeroes in a tremendous gale.'[72] 'They struck on the Faeroe Islands.'[73] 'The ship *Shankinn* was lost in spring with the drift-ice off the Austfjörd.'[74] 'A ship was lost with all her crew that had driven on the Hornstrandir in a heavy gale on the last day of June.'[75] 'The *Stedia Kollu* was wrecked over in the west by Strandir.'[76] Frequently a craft would run on an island, or outlying reef, or shoal. 'A *knörr* struck on the Vestmannaeyjar.'[77] 'Another vessel standing in to the land was wrecked on an island.'[78] 'He stood to the north round Cape Stad, ran on a reef, and lost the ship.'[79] 'Gísli and Vésteinn were at sea for over a hundred days and sailed in the Winter Nights on a rock.'[80] 'The ship broke up on Thjórsársand.'[81] Severe ice conditions obtained on the Iceland coast during several winters in the thirteenth and fourteenth centuries.[82] Many ships are recorded to have been wrecked on the ice-fields. Finally, a considerable number of Iceland-bound craft were forced off their course and driven on to the Greenland coast. Several such cases are recorded during the first half of the 1380s.[83] An interesting point about these shipwrecks is that though on many occasions the ship itself was lost, the crew—and sometimes all or part of the lading—were very often saved.[84] On other occasions, however, vessels went ashore with a heavy loss of life.[85]

The year 1262 marked a turning-point in the history of Iceland. Down to that time Iceland had been a free and independent community, governed, or misgoverned, by her own magnates. The various attempts of the kings of Norway to gain a footing there had been jealously and successfully resisted for nearly four hundred years.[86] On more than one occasion Iceland was preserved from overseas invasion through the formidable navigational difficulties of the project. This autonomy now came to an end, following on—and largely in consequence of—a long period of internal disorder.[81] The Icelanders submitted to the king's authority and agreed to pay him stated taxes, in consideration of which he was to allow them to live in peace under their own laws and to send out six ships annually from Norway to Iceland during the next two summers. This last most important provision was afterwards extended.[88] The new arrangements were by no means wholly satisfactory to the Icelanders, who protested against the policy pursued by the king's government of haling Icelanders to Norway for trial, and who complained that the clause relating to the dispatch of trading-ships to Iceland was not being fulfilled.

When in 1299 King Hákon the Younger succeeded his brother Eirík the Icelanders, before taking the oath of allegiance, demanded redress of grievances. A few years afterwards, in 1303, they reiterated these demands, declaring, 'We will also keep our oath to the kingdom on the conditions agreed to by the king and thanes who inhabit our land. . . . But the promises were that the *sýslumenn* and lawmen in our country should be Icelandic, and that, unless lawfully hindered, six ships should sail from Norway to Iceland every year, laden with such goods as would be of benefit to our country and to us'; and in a memorial addressed to the Norwegian government in 1320 they protested that the arrangement had been that, 'in return for our taxes the king should send hither six ships every summer laden with useful articles, two to the north, two to the south, one to the east and one to the west districts, and that the lawmen and *sýslumenn* and all magistrates in the country shall be Icelanders.'[89] These recurrent protests that the promise about sending the trading-ships to Iceland was not being kept shows how dependent the land was upon supplies procured from overseas. Though it is difficult to know exactly how far it was due to these protests, and how far to other causes, it is apparent that, as the century advanced, there was a marked improvement in communications between Norway and Iceland.[90]

The traffic with Iceland—constituting by far the busiest and most important of all the ocean trade-routes of the North—may be said to have reached its peak just before the middle of the fourteenth century. According to the Icelandic annals, in the year 1340 eleven vessels sailed to Iceland: in 1341, there were six lying in Hvalfjörd alone: in 1345, eleven ships made the Iceland voyage, and twelve in 1346: in 1347, eighteen ships sailed to Iceland and wintered there, besides two others that were wrecked that summer.[91] The shipping engaged in the Iceland traffic belonged partly to the king, partly to the Archbishop of Nidarós,[92] partly to certain cathedral chapters, partly to various magnates, and partly to certain merchants of Trondheim and Bergen.[93] The headquaters of the Iceland trade was now Bergen, which became more and more the market for *skreið*. Since 1294 the staple had been at Bergen, and it was forbidden to foreigners to trade north of that town or with any of the Norwegian dependencies; in the fourteenth century Trondheim merchants who traded to Iceland in stockfish always sold their wares in Bergen.[94] For a brief period in the early 1300s the foreign trade of Norway, as the English customs accounts reveal, was still fairly prosperous. In the next decade or so, however, it is possible to mark a decline. The recession was gradual rather than sudden. But the decline continued throughout the century; and practically the whole of Norway's once flourishing foreign trade passed into alien hands. The question, therefore, arises: why, when Norwegian foreign trade was thus visibly waning, did the ocean traffic with Iceland continue to expand?

Three main factors contributed to this result. First, the merchants of the German Hanse in capturing so much of the Norwegian foreign trade in the Baltic and North Sea had caused a substantial percentage of this trade to be diverted to other routes, thus producing a temporary boom in the Icelandic traffic.[95] Second, the expansion of this ocean trade was closely connected with the growing demand for *skreið*, a commodity which came partly from the Lofoten Isles, and partly (to an increasing extent) from Iceland. Lastly, there is evidence of a notable advance in practical navigation about this time. According to Hauk Erlendsson's recension of the *Landnámabók*, the *leiðarsteinn*, or magnet used as a compass, came into general use in northern waters at some time in the thirteenth century. Rather earlier than this mention is made of the use of the sounding-line for ascertaining the depth and nature of the bottom. Sailing directions for the northern sea-routes made their first appearance in the thirteenth century *Sturlubók* and the fourteenth century *Hauksbók*. The combined effect of these innovations cannot but have served to stimulate the trade with Iceland. There is, however, no evidence of any corresponding advance in the design and rigging of ships. The Icelandic illuminated manuscripts of this period show only the one-masted vessel of the old type, with its one square-sail (with plenty of 'belly' to it). While shipbuilding in England in the later Middle Ages continued to progress, in Norway, comparatively speaking, it was in a state of stagnation.

In the fourteenth century a succession of unparalleled misfortunes overtook the population of Iceland. The year 1300 saw another eruption of the great volcano Hekla;[96] several other eruptions, accompanied by earthquakes, followed in the course of the century. The island was visited by smallpox and other epidemics. A large part of the cattle perished through hunger and disease. During all these afflictions the people of Iceland were to some extent sustained by the trade with Norway. But in the latter half of the century that also failed them. Schreiner has shown that the basic cause of Norway's decline in the fourteenth century was not so much the Black Death, as the economic situation in the North which prevented Norway from recovering from the heavy loss of population occasioned by the plague. Nevertheless, the consequences of the Black Death must not be underrated.[97] The effects of this visitation on the commerce and shipping of Norway were immediately apparent. In 1347 twenty ships or more had wintered in Iceland. But in 1349 there was only one sailing to Iceland: in 1350, and again in 1335, no sailing at all.[98]

Towards the close of the century a partial revival of the trade with Iceland set in, and the dues paid on Icelandic imports constituted an important source of revenue. The *sekkjajald*, which was the duty levied on Icelandic produce imported by Norway, was first imposed in 1382, during the reign of Ólaf Hákonarson. 'The king claimed one-fourth interest in every ship which sailed to Iceland, also 6 out of every 120 fish, and 5 per cent of every roll of *vaðmal* and of every barrel of sulphur and train-oil. This tax was collected both from Iceland and from Norwegian merchants and foreign traders.'[99] In 1375 nine ships were bound for Iceland and six actually arrived. In the winter of 1381 there were ten ships laid up in the Hvalfjörd. In 1389 eleven ships came to Iceland; in the following year, seven; in 1391, eleven. But the revival was short-lived. In 1394 only two arrivals are reported.[100] After the turn of the century things became still worse. The eclipse of her Atlantic trade and shipping was a natural concomitant of the gradual shifting of the centre of gravity from the west coast of Norway to the other end of the kingdom, which was reflected in the transference of the governmental machine from Bergen to Oslo.[101] Communications between Norway and Iceland steadily declined, and early in the second decade of the fifteenth century only the providential arrival of merchantmen from England saved the Icelanders from a fate similar to that of the Norse colonists in Greenland.

14

The Greenland trade-route

THE COLONY OF Greenland was almost, but never entirely, self-sufficing. Their flocks and herds were considerable, and in most years there was an abundant supply of fish. A limited quantity of grain was grown on some of the larger farms;[1] but the majority of the people, as is apparent from the *Konungs Skuggsjá*, had neither tasted nor seen bread.

Owing to the chronic lack of iron in the colony, the Greenlanders had recourse to various ingenious substitutes for metal objects. Thus among the Greenland finds of recent years are an arrow-head, an axe, sledge-runners, and even a padlock of whale-bone. Draughtsmen and chessmen were made out of wood or walrus-ivory for the chequer-board games in which the Scandinavians delighted. Many kinds of farm and household utensils were carved out of the local soapstone. Apart from driftwood from the forests of Siberia, which was carried by the current across the Polar Sea, southward down the east coast of Greenland, thence round Cape Farewell, and up the west coast past the East and West Settlements, the Greenlanders had to depend upon timber fetched from overseas—not only from Europe, but also, in all likelihood, from North America.[2] To what extent they availed themselves of this latter source of supply is uncertain. Iron and iron goods had likewise to be imported from Europe.[3] From the foundation of the Settlements the welfare of the colony had depended upon sea-communications. The trade with Europe was in the long run essential to its prosperity. The position is thus summed up in the *Konungs Skuggsjá*:

But in Greenland it is this way, as you probably know, that whatever comes from other lands is high in price, for this land lies so distant from other countries that men seldom visit it. And everything that is needed to improve the land must be purchased abroad, both iron and all the timber used in building houses. In return for their wares the merchants bring back the following products: buckskins, or hides, sealskins, and rope of the kind that we talked about earlier which is called 'leather rope' and is cut from the fish called walrus, and also the teeth of the walrus.[4]

In the early summer hunting expeditions set out for the uninhabited regions of the North, away beyond the West Settlement: a tract known as the Nordsetr. According to Björn Jónsson, all the large farmers in Greenland had good-sized craft built for these Nordsetr ventures; and sometimes they themselves went with them.[5] The Nordsetr played a highly important role in Greenland's economy. There were found the reindeer, bear, walrus, narwhal, and various species of whale and seal whose flesh, blubber, hides, and ivory were in considerable demand, not only for home use, but also for export to Europe. (It was presumably in large measure in connection with the produce of the Nordsetr that the two *hafskip* belonging to Kolbein Thorljotsson and Hermund and Thorgils Kodransson visited the West Settlement in 1126.[6]) Driftwood, too, came ashore in immense quantities to the north of Disco Bay. The Greenlanders believed that it came from 'the bays of Markland':[7] but in point of fact, as has already been said, it came across the Arctic Ocean from northern Russia. To reach these distant hunting-grounds meant a long and arduous voyage from the Settlements. The Greenlanders built huts and booths for themselves in certain parts of the Nordsetr where they would spend the summer. Early in the fourteenth century they penetrated at least as far north as the 73rd parallel, as was shown by a small stone (now lost) incribed with runes discovered on the island of Kingigtorssuaq. Half a century before an exploring expedition is believed to have reached 76° N.[8] It was in the Nordsetr that the Greenlanders apparently traded with the Eskimos as the Norwegians had trafficked with the Finns.

By far the most valuable export of Greenland was ivory. In the twelfth and thirteen centuries there was a steady demand for tusk-ivory, which for a while practically supplanted real ivory. Whole shiploads of walrus- and narwhal-ivory were imported into Europe from Greenland. In 1266 a ship belonging to Bishop Ólaf left Gardar with a rich lading of walrus-ivory; she went aground at Hitarnes in Iceland; and it is said that centuries afterwards these tusks were still being found on the shore.[9] From a letter of Pope Martin IV, dated 4 March 1282, we learn that the Greenlanders were accustomed to pay their tithes, not in coin, but in sealskins, hides, and ivory.[10] In 1327 the papal legate received from the diocese of Gardar some 250 tusks for Crusade-tithes and Peter's Pence.[11] In tusk ivory Greenland, unlike Iceland, possessed a luxury product. As Dr. Aa. Roussell has observed, 'It is hardly too much to say that the entire trade between Greenland and Europe has been based upon the strong white tusks of the walrus.'

Walrus-rope, or *svarðreip*, large quantities of which were sold, in the thirteenth century, at the great fair of Köln, was held in high esteem for its immense natural strength. A ship's rigging and anchor-cable were commonly made of *svarðreip*.

In the heyday of falconry, the Greenland falcon (*F. candicans*), the largest of the genus, was sometimes exported to the Continent. 'There are also many large hawks in the land,' remarks the *Konungs Skuggsjá*, 'which in other countries would be counted very precious,—white falcons, and they are more numerous there than in any other countries; but the natives do not know how to make any use of them.'[12] At late as 1396 mention is made of twelve Greenland falcons being paid as a ransom to the Saracens for a son of the Duke of Burgundy. It is said that the Greenland falcon was especially prized by sporting prelates.[13]

Occasionally, too, a live polar bear was carried some 2,000 miles across the ocean as a present for a king or some great man in Europe. In the *Hungrvaka* it is stated that Bishop Ísleif gave the Emperor Henry III a polar bear which had come from Greenland, and this bear was 'the greatest treasure'. In 1123 Einar Sokkason shipped a bear, along with his cargo of walrus-hides and ivory, which, as has already been explained, in due course he presented to the King of Norway, in order to gain the latter's support for the establishment of a bishopric in Greenland.[14]

It is impossible to arrive at anything like a precise estimate of the comparative regularity of communications with Greenland; the statistics are inadequate and there are too many unknown factors involved; but it is probable that these communications were rather more frequent than was formerly supposed. There can be no question that only a fraction of the passages to and from Greenland were ever recorded in writing. In the *Sturlubók*, for instance, reference is occasionally made to voyages which find no mention at all in the Icelandic annals. It is also worth noting that what was unquestionably the longest and most formidable ocean venture known to the medieval world was commemorated in a brief sentence or so. 'Afterwards Bárd put to sea and made a fast passage. He came to Greenland.' 'Bárd now put to sea and after a prosperous voyage made a good landfall.' 'It is said of Bárd's voyage that it was a prosperous one. He made a good landfall in Greenland.' 'Thorgrím sailed for Greenland and made a fast passage.' 'Skúf and Bjarni put to sea. They got a fair wind and had a prosperous voyage and came to Norway.' 'The following summer they sailed for Norway and made a fast passage.'[15] Further, though it is certain that there was sometimes an interval of several years between one Greenland sailing and the next, it is to be remembered that there might sometimes be several ships arriving in Greenland in a single season. This had happened in the early years of the colony, and it happened again in 1131.[16]

Some index to the regularity or otherwise of communications with Norway is furnished by the following list of the bishops of Gardar, which covers the greater part of the twelfth, thirteenth, and fourteenth centuries.

Arnald	Consecrated in 1124; arrived in Iceland in 1125; reached Greenland in 1126; returned to Norway in 1150.
Jón Knút	Consecrated in 1150; died in 1187.
Jón Smyrill	Consecrated in 1188; reached Greenland in 1190; made several visits to Iceland; thence sailed to Norway and journeyed to Rome; died in 1209.
Helgi	Reached Greenland in 1212; died in 1230.
Nicholas	Consecrated in 1234; reached Greenland in 1239; died in 1242.
Ólaf	Consecrated in 1246; reached Greenland in 1247; stranded on the Iceland coast in 1261; left Iceland for Norway in 1264; returned to Greenland in 1271; died in 1280.
Thord	Consecrated in 1288; reached Greenland in 1289; left Greenland, to which he never returned, in 1309; died in 1314.
Arni	Consecrated in 1314; reached Greenland in 1315; died in 1351.
Jón Skalli Eiríksson	Succeeded Arni in 1349; never reached Greenland; died in 1351.
Álf	Consecrated in 1365; reached Greenland in 1368; died in 1378.

At the same time the Icelandic annals and sagas record a grisly toll of shipwrecks, strandings, and crews overtaken by *hafvilla*, craft headed off and hopelessly adrift on the high sea.[17] From the early years of the colony this voyage had been regarded by experienced mariners as a difficult and dangerous one. When Greenland was settled, it will be remembered, of the twenty-five vessels which sailed from Iceland, only fourteen ever reached their destination. In the reign of Ólaf the Saint a widely travelled Icelander, Thórarinn Nefjólfsson, expressly warned the king that a vessel might well fail to reach Greenland at all, as had frequently happened in the past.[18] It is on record that when one Ketil 'the Stammerer' was blown off his course to

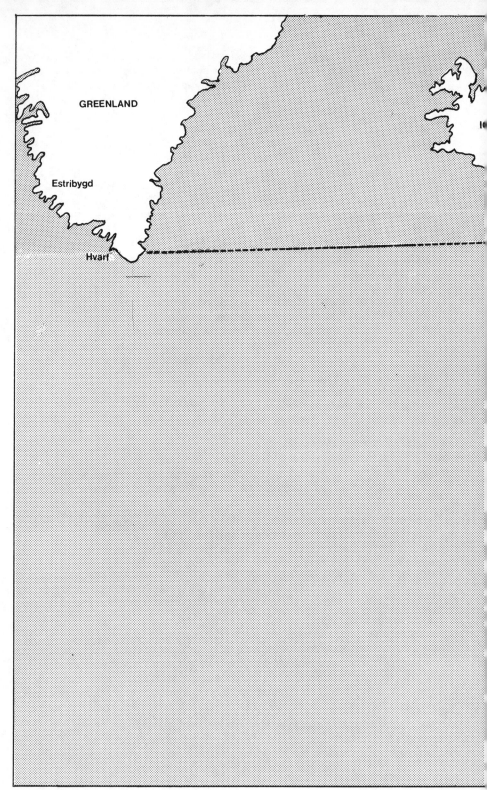

The course for Greenland

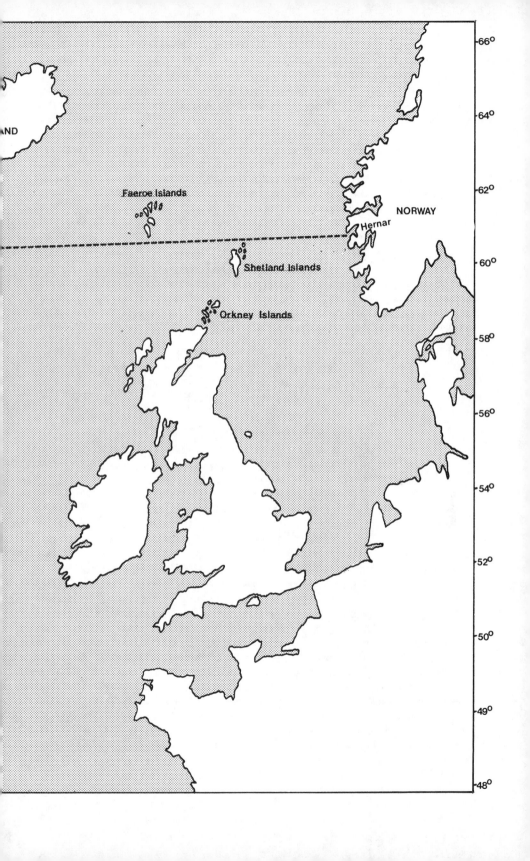

Greenland in 1208, he did not return to Iceland until four years afterwards.[19] Of the three vessels which in 1125 sailed from Norway for Greenland, only one ever arrived in the Eiríksfjörd.[20] A few years later there was the same high proportion of losses in a single season. In one decade of the following century, the 1260s, the Greenland *knörr* was twice wrecked—in 1260 and 1266.[21] On several occasions in the 1360s the merchantmen which kept up communication between Norway and Greenland were lost.[22] According to the *Konungs Skuggsjá*, two special perils of the Greenland Sea were the vast ice-fields extending all the way down the East Greenland coast to Cape Farewell, which were the cause of many unrecorded wrecks, and the dread *hafgerðingar*, or 'sea-hedges', which, mentioned in the earliest times of the colony, appear to have been occasioned by severe sea-quakes. Of the ice more will be said later. The *hafgerðingar* are described in the *Konungs Skuggsjá* as follows:

Now there is still another marvel in the seas of Greenland, the facts of which I do not know precisely. It is called 'sea-hedges', and it has the appearance as if all the waves and tempests of the ocean have been collected into three heaps, out of which three billows have formed. These hedge in the entire sea, so that no opening can be seen anywhere; they are higher than lofty mountains and resemble steep, overhanging cliffs. In a few cases only have the men been known to escape who were upon the sea when such a thing occurred.[23]

Apart from the voyages which terminated in total disaster, with the loss of both vessel and crew, there were a good many others at the end of which, though the vessel itself was lost, the crew managed to struggle ashore to safety.[24] It is not surprising that throughout the Middle Ages the navigation to Greenland was regarded as the most hazardous venture in the northern seas.[25]

Yet considering the manifold hazards of the Greenland trade-route, what is chiefly remarkable is that, despite severe losses, the annals record the safe arrival of so many vessels that had made the long ocean crossing. It is worth noting that during the twelfth, thirteenth, and fourteenth centuries successive Bishops of Gardar journeyed to and fro between their distant diocese and Iceland and Europe. None of these prelates was ever lost at sea.

Notwithstanding a general belief to the contrary, these ocean voyages were *not* made by longships.

This error continues to be propagated, unsupported by any kind of evidence. For example, H. A. L. Fisher wrote that 'the long Viking warships, driven by oar and sail, planted colonies on the bleak shores of Greenland and, six hundred years before the voyage of Columbus, explored the North American coast'. Similarly, E. M. Carus Wilson declared that 'big timber for the building of churches was laboriously brought in long ships across the ocean to Iceland'. F. W. Brooks fell into the same error, as did P. K. Kemp, J. W. Thompson, and several others.[26]

On the contrary, it was in a merchantman (*kaupskip*), according to the *Grœlendinga saga*, that Thorfinn Karlsefni sailed southward down the American coast. It was likewise in a *kaupskip* that Einar Sokkason came to Norway with his cargo of walrus ivory and a live polar bear. In the *Grœlendinga pattr* the ship belonging to the merchant Arnbjörn is termed *mikit hafskip*. The Greenlandman which, under the command of Grimar and Sörli, arrived in the Vestmannaeyjar in 1218, is described as a *knörr mikil*. It was a *knörr* which carried the Bishop of Greenland to his distant diocese.[27] There is no record of longships sailing to

Greenland or North America, or even (with one exception) to Iceland[28] or the Faeroes.

This may well have been due partly to political and military reasons: from their fastnesses in the Orkney and Shetland Islands, as has already been said, recalcitrant chieftains had been able to harass the coasts of Norway. Iceland and Greenland, and even the Faeroes, however, were too far off to constitute a serious threat. But the main and essential reason why it was the *hafskip* and not the *langskip* that made the long ocean crossing to the distant overseas settlements of Norway was, as is clear from the evidence of the sagas, that the *hafskip* was adapted to ocean navigation and the *langskip* was not.[29] From a certain passage in the *Ólafs saga helga* we learn that it was exceptional for a longship to have as much freeboard as a *hafskip*.[30] It is significant, too, that the Icelanders refused to let Ólaf the Saint get possession of Grímsey, fearing that that island would be used as a base for 'a great war-fleet cruising from thence in longships'. It was said that the sea between Norway and Iceland was so wide that longships could not make the passage. Even the much shorter ocean crossing between Norway and the Faeroe Islands was believed to be too much for the long, slender *langskip*, with its low freeboard and comparatively frail construction. 'Longships cannot get there on account of the gales and tidal streams, which are often found to be so strong that a merchantman can scarcely bear up against them.'[31] The Viking *langskip*—which was essentially a fighting ship, a war-galley 'shield-hung from stem to stern', with its menacing dragonhead mounted at the stem, and golden wind-vane glittering at the masthead—sailed to and fro across the Baltic, the North Sea, and the Channel: but it had never crossed the Western Ocean.

Much has been heard of the longship which, as has been said, was the chief and fundamental factor in Viking raids and expansion on the Continent. Not nearly so much of the *hafskip*, or *knörr*: though this, it must be emphasized, represented, in complete mastery of technique and design, one of the most interesting and important achievements of the medieval era. The evolution of the *knörr* had in fact made possible the settlement of the Faeroe Islands, Iceland, and Greenland and the inaguration of the earliest ocean trade-routes in northern waters. For the first time in history, outlying regions of the western hemisphere had been brought into the European trading-system.

For centuries the navigation to Greenland was based on oral tradition and experience. The sailing directions were handed down from generation to generation; they were rarely committed to writing, and even then—in all likelihood—only in part: in their oral form they would most certainly have been a great deal fuller. The course for Greenland in the later medieval period is recorded in some detail in the fourteenth-century *Hauksbók*. 'From Hernar in Norway sail due west for Hvarf in Greenland; and then will you sail north of Shetland so that you can just sight it in clear weather; but south of the Faeroe Islands, so that the sea appears half-way up the mountain-slopes; but steer south of Iceland so that you may have birds and whales therefrom.'[32] The outward-bound *knerrir* would cross the ocean in approximately lat. 61° N: a course which would take them north of the Shetlands, and south of the Faeroes, at about the distances prescribed.[33] Still holding that course they would pass some 150 miles to the south of Reykjanes, which is about the distance at which birds from Iceland are first sighted by modern Icelandic trawlers. The crossing and recrossing of this immense span of open sea—more than 1,000 nautical miles—between the Faeroe Islands and a point on the East Greenland coast about 60 miles north of Cape Farewell may justly be regarded as the supreme achievement of Norse seamanship and navigation. The small, sturdy *knerrir*, with their single mast and one square-sail, rising and falling to the scend of the long western seas further from sight of land than any other craft of their day, constituted the one frail tenuous link which bound this far colony of Greenland, for four long centuries, to Christianity and Europe.

The rest of the voyage was essentially a matter of coastal navigation. The course for the inhabited regions of the country lay southward down the edge of the ice-floes, and around Cape Farewell.[34] Here, at the termination of the long sea-route from Norway, rose up the bold massif at the southernmost end of Greenland, 'the high land of Hvarf', as the mariners called it. Here, too, under Herjólfsnes lay the first regular port of call, the 'harbour which is called Sand-haven, where merchants from Norway were wont to come'.[35]

In these *knerrir*, then, came the imports from Europe of which Greenland stood so sorely in need: iron and iron goods, tar, timber, malt, and grain. On the homeward passage they carried back to Norway the produce of the colony: walrus and narwhal tusks, sealskins and other hides, walrus-rope, wadmal, and falcons.

Greenland, as we know, was the westernmost land in which the Norse stock was able to establish itself. It also marked the limit of their navigation for direct ocean passages. Whether the *knörr* and the navigation of the Norsemen would have been equal to the voyage from North America to Europe will never be known, for they were never put to the test.

The headquarters in Norway of the Greenland trade, in the Viking era, was the port of Nidarós, or Trondheim. Nidarós had received its charter from Ólaf Tryggvason in 998: it was for a time the most important trading town in Norway and the chief centre of her overseas trade. But during the thirteenth century Bergen became a formidable rival to Trondheim, and in 1294 was made the staple town for all traffic with the Norwegian dependencies, including Greenland. Already by about 1250 Bergen was beyond comparison the greatest centre of trade and shipping in Norway. It is recorded that Greenlanders were among the cosmo-politan assemblage—Icelanders, Englishmen, Germans, Danes, Swedes, and Gotlanders—that thronged the quays of Bergen.[36] Proof of this lies in the letter which in 1325 Bishop Audfinn of Bergen addressed to the Archbishop of Nidarós, complaining that 'the merchants of Trondheim who now come in *knerrir* from Greenland' were refusing to pay their tithes on these cargoes shipped to Bergen into the hands of the chapter there.[37] Fifteen years later, there arose a similar dissension between the Archbishop of Nidarós and the Iceland traders in the town.[38] It would appear that the trade to both Iceland and Greenland was in the hands of a few families of Trondheim. In the letter cited above two of these merchants are mentioned by name: Ólaf in Lexo and Aendrida Arnason. The Greenland merchants seem to have been joined together in some kind of guild or trade association.[39]

It was the policy of Hákon Hákonarson to make the Atlantic colonies of Norway directly subject to the Crown. After prolonged negotiations, and with the powerful support of the Church, this was finally accomplished at the end of the reign, during the years 1261–4. Henceforward the Greenlanders, as well as the Icelanders, acknowledged the royal supremacy. They accepted the obligation of paying tax and further agreed to pay fines to the king for manslaughter, whether it were Norwegian or Greenlander that was slain, and whether the homicide were committed in the inhabited part of the country or in the Nordsetr. The king's authority, it was proudly claimed, now extended to beneath the very Pole-star.[40]

There can be little doubt, however, that Greenland had become economically dependent upon Norway long before this. Only in the early days of the colony had the Greenlanders possessed their own *hafskip*. The working life of these vessels was only thirty years or so, and the Greenlanders had no timber of the own for shipbuilding. In 1189, according to the Icelandic annals, there came from Greenland to one of the havens of Iceland a ship 'held together with wooden pegs and the sinews of animals'.[41] This does not sound like a *knörr*. Long before the formal union of Norway and Greenland, traffic between the two countries has passed altogether into Norwegian hands. By the act of union the king is believed to have assumed responsibility for the provision of shipping. It would seem that part of the agreement was that the king should dispatch one ship annually to Greenland, in the same way that he had promised to send six ships to Iceland. If this was indeed the case, the agreement came too late to be of much benefit to the Greenlanders. For with the reign of Hákon Hákonarson the great days of medieval Norway ended. Under his son and successor, Magnus, the decline set in. From now on the trade and shipping of Norway were progressively undermined by the enterprise and aggression of the German Hanse.

Moreover, the course for Greenland had had to be changed on account of the expansion of the ice-fields, which rendered impracticable the track set out in the old sailing directions; and significant allusions to the rigours of the Greenland passage seem to suggest that communication with the colony was becoming more difficult and uncertain.[42] It will be recollected how in the early days of the colony it was possible to cross the sea between Snaefellsnes and the East Greenland coast in about the 65th parallel. 'From Snaefellsnes in Iceland to the nearest point of Greenland is four days' sailing across the sea to the westward', ran the sailing directions in the *Sturlubók*.[43] It is significant, too, that in earlier centuries quite a number of vessels are recorded to have been wrecked, not on the edge of the drift-ice, but on the East Greenland coast itself. Early in the history of the colony there is a vivid account of such a shipwreck and of the desperate struggle of Thorgils Orrabeinsfostri and the other survivors to reach the Eystribygd and safety, in the *Flóamanna saga*. It is stated there that Thorgils's ship was wrecked on the coast itself, 'by a sand-dune in a certain bay', just a week before winter.[44] The loss of Arnbjörn's *knörr*, early in the twelfth century, and the later wreck of the *Stangarfoli*, have already been noted.[45]

Towards the thirteenth century, however, the conditions deteriorated. There are significant references to drift-ice off the coasts of Iceland, during the thirteenth and fourteenth centuries, in the Icelandic annals. By about 1200 the ice-floes carried down by the Polar current had increased to such an extent that, according to the *Konungs Skuggsjá*, the old course had to be abandoned.

As soon as he [the mariner] has passed over the deepest part of the ocean, he will encounter such masses of ice in the sea, that I know no equal of it anywhere in all the earth. Sometimes these ice-fields are as flat as if they were frozen on the sea itself. They are about four or five ells thick and extend so far out from the land that it may mean a journey of four days or more to travel across them. There is more ice to the north-east and north of the land than to the south, south-west, and west; consequently, whoever wishes to make the land should sail around it to the south-west and west, till he has come past all those places where ice may be looked for, and approach the land on that side.[66]

From the description of these vast ice-fields in the *Konungs Skuggsjá* it is apparent that ships must now have been wrecked on the ice-edge, rather than on the coast itself.

It has frequently happened that men have sought to make the land too soon and, as a result, have been caught in the ice-floes; and some have perished in them; but others again have got out, and we have seen some of these and have heard their accounts and tales. But all those who have been caught in these ice drifts have adopted the same plan: they have taken their small boats and have dragged them up on the ice with them, and in this way have sought to reach land; but the ship and everything else of value had to be abandoned and was lost. Some have had to spend four days or five upon the ice before reaching land, and some even longer.[47]

Later evidence to very much the same effect is to be found in the description of life and conditions in Greenland compiled by the Norwegian cleric, Ívar Bárdarson, who administered the estates belonging to the see of Gardar in the middle of the fourteenth century. It is to Ívar Bárdarson that we owe a detailed list of the fjords and islands of southern Greenland, and also a most interesting set of sailing directions. He emphasizes the fact that the ice conditions had become immeasurably more severe than they had been in an earlier era.

Item from Snaefellsnes in Iceland, where the distance to Greenland is shortest, two days' and two nights' sailing due west is the course, and there lieth Gunnbjarnarsker half-way between Greenland and Iceland. This was the old course, but now ice has come down from the north-west out of the gulf of the sea so near to the aforesaid skerries, that no one without extreme peril can sail the old course, and be heard of again. . . . Item when one sails from Iceland, one should shape one's course from Snaefellsnes . . . and then sail due west one day and one night, but then slightly south-west to avoid the aforesaid ice, which lieth off Gunnbjarnarsker, and then one day and one night due north-west, and so one comes right under the aforesaid high land of Hvarf in Greenland, under which lieth the aforesaid Herjólfnes and Sandhaven.[48]

The difficulty of transacting business with Greenland is mentioned in a letter written in 1341 by Bishop Hákon of Bergen, who asserted that the way lay 'per mare non minus tempestuosissimum quam longissimum.[49] Communications with Greenland were now becoming very irregular. Moreover, it may be safely said that in no one year in these later centuries is there any record of more than one *knörr* making the long venture to Hvarf. In the middle of the fourteenth century there was actually an interval of nine years between one sailing and the next. Later, on a rumour of Bishop Arni's death reaching Norway, his successor was appointed while the old bishop was in fact alive in Gardar. After Arni's death in 1348 there was no bishop in the land until Álf came out twenty years afterwards.[50]

On the other hand, the effect on navigation of the worsening ice conditions ought not to be exaggerated. The long interval already mentioned—between 1346 and 1355—during which no passage is recorded to have been made to Greenland may very well be explained by the calamitous consequences of the Black Death.[57] These consequences are similarly reflected in the statistics of the Iceland trade. Though her foreign trade was rapidly declining, the ocean traffic between Norway and her dependencies was still vigorous and important in the early half of the fourteenth century. The Bishop of Bergen's protest to the Archbishop of Nidarós over the Greenland tithes, quoted above, suggests that in the first quarter of that century a fairly substantial traffic was still carried on between Greenland and the motherland;[52] and this is borne out by a series of voyages to Greenland listed in the Icelandic annals. It is also to be remembered that a particular sailing was as a rule only recorded when some distinguished person was on board (such as the bishop of Gardar);[53] or when, as in 1346, an unusually rich cargo was carried;[54] or when, as in 1355, there was something remarkable to record;[55] or, again, when the voyage ended in shipwreck.[56] At irregular intervals a *knörr* would be fitted out for the hazardous voyage across the ocean. From time to time shiploads of walrus-ivory would arrive in Norway from Greenland. Successive bishops of Gardar would be duly elected, consecrated (usually in Trondheim cathedral), and, eventually, would take passage for their dioceses 'at the world's end'.[57]

In the latter half of the thirteenth century the Greenland trade was subjected to an increasing measure of governmental control, and at last, in about 1300, was made a royal monopoly. It would seem that by this time the sailing of a *knörr* was sufficiently unusual to be singled out for special mention. The ban on unauthorized trade was rigorously enforced. None might venture to Greenland without leave from the king. In the last part of the fourteenth century, as will appear in a later chapter, it is stated that when a party of Icelanders bound from Norway to Iceland were forced off their course to Greenland—where they were obliged to remain for four winters—they were actually arrested, on their arrival in Norway, for having trafficked with Greenland without the king's permission.[58] The Greenlanders dared not even take advantage of such craft arriving by chance in their country to send off the goods belonging to the Crown which had long been accumulating in the store-houses.[59] In point of fact, the Greenland market had largely lost its importance. Walrus- and narwhal-ivory were no longer much in demand.[60] Russian furs had practically supplanted the hides of Iceland and Greenland. Bergen, the centre of the Greenland trade, had fallen on evil days. Devastated by the Black Death in 1349, its once flourishing trade now for the most part in the hands of the German Hanse, Bergen at this stage could scarcely find shipping for the Iceland, far less the Greenland, passage. In the last part of the fourteenth century allusions to the Greenland *knörr* became increasingly rare. In 1366 a merchantman, the *Bauta hluti*, was fitted out for the Greenland voyage.[61] The *Grænlands knörr* was lost in 1367 or thereabouts, and there is no record of any successor.[62] It is significant, too, that the bishops of Gardar no longer reside in their diocese. Bishop Álf, it will be remembered, had died in 1378. A new bishop was in due course nominated, but he never made the voyage to Greenland. It is said that on the death of Bishop Álf an old priest took over certain of his functions. Though successive bishops of Gardar continued to be appointed down to the time of the Reformation none of these ever visited the diocese.

Regular sailings between Greenland and the Norwegian homeland had now all but ceased. The Icelandic annals recorded that, in 1385, 'a ship arrived in Norway from Greenland. She had wintered there for two years, and on board her were some survivors from the wreck of the *Thorláksúðen*. They brought news of Bishop Álf's death, which had occurred six years before.'[63]

15

Seamanship and navigation

As MIGHT BE expected, in the main a remarkably high standard of seamanship prevailed among the Norse mariners. The crews had confidence in themselves and in the craft they sailed—in their own tough frames and stout hearts, in the instinctive resource and skill born of long years of experience on the coast and on the high sea, in well-found craft and good gear—and the event, for the most part, justified this confidence.

In the early Norwegian laws there are special provisions relating to the seaworthiness and lading of vessels. With certain limits ships must be watertight,[1] and they must not be overloaded.[2] The latter danger was clearly recognized in the twelfth-century *Gulathinglaw*. 'If a man loads a merchant ship at his home in the country and men take passage with him, he shall load the ship in such a way that all who have taken passage with him, will have room. Now if the ship is loaded too heavily, the owner shall return his own wares [to the shore] and those who have taken passage shall have the necessary room. If they [still] think it unseaworthy, those who took passage last shall remove their effects, and each one shall have six oras for breach of contract.' In the autumn ships were hauled on shore by means of ropes and rollers, and laid up for the winter in the shelter of a shed, where they were thoroughly overhauled and freshly tarred. 'When a ship comes home to the beams, it shall be hauled up and all the gear shall be put away.' In these laws there are numerous regulations governing the laying-up of ships (*uppsát*) for the winter.[3]

'If', declares the father in the *Konungs Skuggsjá*, 'you are preparing to carry on trade beyond the seas and you sail your own ship, have it thoroughly coated with tar in the autumn and, if possible, keep it tarred all winter. But if the ship is placed on timbers too late to be coated in the fall, tar it when spring opens and let it dry thoroughly afterwards. Always buy shares in good vessels or in none at all. Keep your ship attractive, for then capable men will join you, and it will be well manned. Be sure to have your ship ready when summer begins and do your travelling while the season is best. Keep reliable tackle on shipboard at all times, and never remain at sea in late autumn, if you can avoid it.'[4]

In the same work it is related how a merchant vessel must be provided with stores and tools for dealing with any repairs that might become necessary in the course of a voyage. Both *hafskip* and *langskip* were liable to all manner of injuries; and these could not always wait over until such time as the vessel arrived in harbour.[5]

'Whenever you travel at sea', the father continues, 'keep on board two or three hundred ells of wadmal of a sort suitable for mending sails, if that should be necessary, a large number of needles, and a supply of thread and cord. . . . You will always need to carry a supply of nails, both spikes and rivets, of such sizes as your ship demands; also good boat hooks and broad-axes, gouges and augers, and all such other tools as ship carpenters make use of. All these things that I have now named you must remember to carry with you on shipboard, whenever you sail on a trading vessel and the ship is your own.'[6]

In the *Fóstbrœðra saga* there is a lively description of a major reparing job being carried out in mid-ocean, the ship being on passage to Greenland. The sail-yard had split in a violent squall and the sail went overboard. The crew seized the sail and managed to haul it in again. Then Skúf, the master, asked two of the crew, Gest and Thormód, to fish the yard for him, which they did with remarkable skill; after which they bent the sail to the yard, and the ship continued her voyage.[7]

When a ship was prepared for sea, the hull and standing and running rigging were carefully overhauled,[8] and ballast was taken on board.[9] After launching it was customary for vessels to lie under some promontory or island, sometimes for weeks on end, while waiting for a fair wind.[10] In the thirteenth-century Icelandic sagas there are many references to the signs of the weather, and a mariner would be described as *veðr-kœnn*, skilful in forecasting the weather. It would appear from these passages that the Norsemen must have possessed this quality in a high degree. Thus Ohthere at the outset of his voyage around the North Cape and along the Murman coast to the White Sea had chosen a time when the meteorological conditions apparently held out the promise of favourable winds for his venture.[11]

In the *Konungs Skuggsjá* we learn what voyages were practicable at what seasons. In answer to his son's inquiry about the proper time of year for putting to sea the father declares that it is impossible to lay down a hard and fast rule, 'for the seas are not all alike, nor are they all of equal extent. Small seas have no great perils, and one may risk crossing them at almost any time; for one has to make sure of fair winds to last a day or two only, which is not difficult for men who understand the weather. . . . But where travel is beset with greater perils, whether because the sea is wide and full of dangerous currents, or because the prow points towards shores where the harbours are rendered insecure by rocks, breakers, shallows, or sand-bars—whenever the situation is such, one needs to use great caution; and no one should venture to travel over such waters when the season is late.'

The father continues: 'Now as to the time . . . it seems to me most correct to say that one should hardly venture over-seas later than the beginning of October. For at that time the sea begins to grow very restless, and the tempests always increase in violence as autumn passes and winter approaches. . . . Men may venture out upon almost any sea except the largest as early as the beginning of April. For at that time when we date 16 March, the days lengthen, the sun rises higher, and the nights grow shorter.'[12] All of which, by the way, is in fairly close accord with the testimony of the sagas and the Icelandic annals.

The *Konungs Skuggsjá* paints a vivid picture of the life of the seafaring trader of those times; it constitutes, indeed, one of our most valuable sources of information about the merchant mariner of the North. It is a fitting memorial of the great days of Hákon Hákonarson: its outlook is consistently matter-of-fact, prudent, and practial: it is refreshingly free from contemporary superstitions: and its unknown author has an agreeable knack of usually putting a matter in a nutshell. Thus: 'The man who is to be a trader will have to brave many perils, sometimes at sea and sometimes in heathen lands, but nearly always among alien peoples; and it must be his constant purpose to act discreetly wherever he happens to be. On the sea he must be alert and fearless. . . .'[13]

He must indeed.

The position of the master, or *stýrimaðr*, to some extent resembled that of the skipper of a small fishing-vessel of today. The crew were under the master's orders; but it was customary for the *stýrimaðr* to consult the others and to listen to their various opinions, before finally making up his mind. It often happened that among the crew were a number of merchants, each with a share in the lading of the ship, who would naturally expect to be consulted on any point of importance.[14] At sea the principal duties of the crew were handling the sail, steering, baling, and standing look-out. As one would expect, some of the crew were on watch, while the others slept or otherwise took their ease.[15]

Among the orders which have come down to us in the sagas are, *Meir á stjórn*, 'Starboard', *Betr á bakborða*, 'Port', and *Halt svá fram*, 'Hold her steady'; also in the *Fornmanna sögur* an interesting fragment of seamen's dialogue has been preserved: 'When they [the vessels] met, a man stood up in the pinnace, tall, of distinguished appearance, and richly apparelled, and inquired: "Who is the master? Where did you spend the winter? Where did you take your departure? Where did you make the land, and where did you pass the night?" Halli replied: "Sigurd is the master; we wintered in Iceland, we sailed from Gásar, we made the land at Hítrar, and passed the night under Agdanes".'[16]

In early times, at any rate, the different members of the crew were more or less on an equal footing. Thus we read in the *Eyrbyggja saga*: 'Thorleif Kimbi took ship that same summer with traders who got ready in Straumfjord, and was a messmate of the master's. In those days it was the custom of traders to have no cooks, but the messmates chose by lot among themselves who should have the duty of the mess day by day. Then too it was the custom of all the seamen to have their drink in common, and a cask should stand by the mast with the drink therein, and a locked lid was over it. But some of the drink was in tuns, and was added to the cask thence as soon as it was drunk out.'[17] The usual shipboard fare was meal and butter, also *skreið*, burgoo, and bread. The drink as a rule was water, and occasionally whey or beer was taken.[18]

During the reign of Magnus Hákonarson a system of fines was instituted for seamen who broke their contracts. If this happened in Denmark, Gotaland, or Sweden, the offender was mulcted of two silver marks—half to be paid to the king, half to the master of the craft; if in Gotland or Småland, four silver marks; if in England, the Orkneys, Shetlands, or the Faeroes, eight silver marks; if in Greenland, Iceland, or Russia, two and two-thirds ounces and thirteen silver marks.[19]

In the last part of the twelfth century, as is clear from the *Gudmundar saga*, the *vindáss* was in use in the larger craft of Scandinavia for raising the sail-yard. It consisted of a round horizontal beam turned between two bitts by a handle. We may observe such a *vindáss* in use in the Winchelsea seal (*c.* 1285), which shows the anchor-rope being hove in by means of a windlass mounted abaft the mast. In the fifteenth-century Kalmar ship the *vindáss* is situated in exactly the same position as the craft depicted in the Winchelsea seal.[20]

The art of sailing on a beam wind, and even of tacking, can be traced at least as far back as the Viking age. The blocks for securing the *beitiáss* have been found in the Gokstad vessel. But though it is evident that the Norse *hafskip* and *langskip* could sail close-hauled, it would appear that they did so only over comparatively limited distances, for example, for making an Iceland haven.[21] There is no record of ships sailing by the wind for long periods. The usual procedure, as is clear from a good many passages in the sagas, was for the crew to anchor and wait for a fair wind. On the high sea, when the wind hauled ahead, and blew a dead muzzler, the crew simply allowed the vessel to drift. Thus in the *Guðmundar saga*: 'Then again they met an east wind and drifted before it once more, and their ship was driven west out to sea.'[22] But it the sea ran too high, they had to bring the ship's head to it. The paramount importance of keeping the ship head on to the seas is also alluded to in the *Guðmundar saga*. 'Then they drifted once more at the mercy of the gale, and during the night the men who were awake keeping watch heard loud and terrible crashes, and saw a huge roller which they thought would be their destruction if it struck the ship broadside on. The steersman wanted to hoist sail, saying that they could only save their lives if they managed to turn the prow of the ship by this means.'[23]

In bad weather they would reduce sail by taking in one or more reefs. Their method of doing so was to lower the yard, take in the reef, and then hoist the yard up again. In severe gales the sail was sometimes goosewinged.[24] It is recorded in the *Guðmundar saga* that 'the gale was so strong that they sailed on one reef';[25] and in the *Orkneyinga saga* that 'when they drew out from the Isles the wind began to freshen. Then it blew so strong that they had to take in sail in the smaller ships.'[26]

In gales they sometimes mounted weather-boards (*vígi*) on the sides above the bulwarks.[27] Sometimes the hull had to be strengthened by 'undergirding' by means of ropes called *þvergyrðingar*.[28] In an emergency they might also lighten the ship by jettisoning part of her lading, as is recorded in the *Jónsbók*.[29] Occasionally, to save the ship, the mast might be cut down.[30]

Pumps were not used on shipboard until after the Viking age. In the *Grettis saga* it is clearly stated that, in the time of Grettir, there were no pumps fitted in ocean-going craft.[31] The old method of baling was with bilge-buckets, as described in the *Fóstbræðra saga*. Two men were told off for the baling; one stood below in the baling-well and the other on deck. The former filled the bucket and handed it up to his mate; and the latter hauled it to the gunnel and emptied it over the side.[32] There is a reference to the 'baling-watch' in the early Norwegian laws.[33] Mention is made of ship's pumps in the thirteenth-century sagas.[34]

It sometimes happened that when a ship was set on a lee shore, her crew would steer straight for the land in the hope of saving life, even if ship and cargo had to be sacrificed. On many occasions it is said that though the craft, and sometimes the whole of her lading, was lost, her people came safe and sound to land.[35] The hazardous operation of running in under sail through a heavy sea (*sigla til brots*) seems to have been accomplished with great skill. In the danger area of a surf the slightest error of judgment would probably cause the vessel to broach to, or slew round broadside on in the trough of a sea. The buoyancy and elasticity of these Norse sailing craft—resulting from the peculiar method of construction already remarked upon—served to make this manœuvre possible. There is a good example of running a vessel ashore in the *Orkneyinga saga*. 'Wednesday was very stormy, but during Thursday night they sighted land. It was then very dark. They could see breakers all around them. Up to this time they had held together. Now there was no choice but to run the two ships ashore; and so they did. There was a rocky beach in front of them, and only a narrow foreshore and cliffs beyond. Then all the men were saved, but they lost much of their belongings; some of them were thrown up during the night.'[36] Another example of *sigla til brots* may be quoted from the *Egils saga*. 'Then they sailed southwards past Scotland, and had great storms and cross winds. Weathering the Scottish coast they held on southwards along England; but on the evening of a day, as darkness came on, it blew a gale. Before they were aware, breakers were both seaward and ahead. There was nothing for it, but to make for land, and this they did. Under sail they ran ashore, and came to land at Humbermouth. All the men were saved, and most of the cargo, but as for the ship, that was broken to pieces.'[37]

There are significant passages in the sagas which reveal what careful attention the Norsemen paid to the set and drift of currents and tidal streams. Under certain conditions of wind and tide they were evidently unable to sail against the tide; it was their practice on these occasions to anchor and wait for it to turn. In the *Egils saga* we hear how Grím the Hálogalander and his crew sailed up the Borgarfjörd past all the skerries and then anchored until the wind fell and the weather cleared. They waited for the flood and then brought the vessel into an estuary called Gufuá.[38] Similarly, in the *Orkneyinga saga* it is told how Eyvind sailed from Ireland, bound for Norway. But as the wind was fresh and the tide against him Eyvind turned into Ásmundarvág in the Orkneys, and lay there for some time weather-bound.[39] The Norsemen were accustomed to treat such races as Pentland Firth and Dynröst with considerable respect.[40]

In the sagas there are numerous passages which show how cautiously mariners approached the land. There is a typical example of this in the *Egils saga*.

They had a strong breeze, and were but a little time out ere they came to the south coast of Iceland. The wind was blowing on the land; then it bore them westwards along the coast, and so out to sea. But when they got a shift of wind back again, then they sailed for the land. There was not a single man on board who had been in Iceland before. They sailed into a wondrous large firth, the wind bearing them towards its western shore. Landwards nothing was seen but breakers and a harbourless shore. Then they stood slant-wise across the wind as they might (but still eastwards), till a firth lay over against them, into which they sailed, till all the skerries and the surf were passed. Then they put in by a ness. An island lay out opposite this, and a deep sound was between them: there they made fast the ship.[41]

There is a passage worth quoting in the *Laxdœla saga* which tells how an Icelandic *hafskip* was brought into a safe anchorage on a strange coast.

So they sailed on, night and day, but ever with light winds. Until on a certain night the men on watch jumped up and called all hands to turn out in a hurry; saying they could see land ahead and so close that they were near running aground. The sail was set, with a very light wind. The men turned out at once, and Örn gave orders to stand off the shore if it could be done. But Ólaf said: 'We can do nothing of the kind. I can see the surf on both bows and all about astern. Take in sail, and hurry. Then we will see what is to be done, when daylight comes.' None of them knew what land it was. So now they put out an anchor, which caught ground immediately. There was much talk as to where they had come to; but as soon as daylight came they knew it for Ireland. Then said Örn: 'I am thinking that we have not come into a right good berth. For this place is far from any port and from those towns where outlanders have the right of entry. Also we are left high and dry by the ebb, like a stranded

stickleback. . . . But I have noticed that there has been a gathering together of folk back up
on land today, and the Irish seem to be concerned about the ship's coming ashore. Also I
noticed today when the tide was out that there is a creek running in behind this headland, a
river-mouth, and that out of this outlet the sea runs smooth and free all the while. Now, if
our ship is not damaged, then we can put out our boat and tow our ship into the place.'
There was a clay bottom where they had been lying at anchor, and no plank had been sprung
in their ship. Ólaf's crew shifted the ship over and cast anchor.[42]

Another example of this kind of skill, the product of long years of seafaring
experience, is to be met with in the *Guðmundar saga*.

One evening they sighted land and drew so near to it that they had the breakers on both
sides. They realized that they had driven back to the Hebrides, but there was nobody able to
steer the course in those waters, and most of them expected that the ship would be dashed to
pieces and those on board perish. While they were in such great peril the merchants could
think of no way out, and then the Bishop-elect spoke to Hrafn, and asked him to act as pilot.
Hrafn wished to be excused from this, saying that he had never been in those waters before,
but Gudmund told him to try, and said that such luck as his could accomplish much. Hrafn
answered: 'The master's word is law', and asked Gudmund to give him his blessing. Then he
took over the navigation of the ship with the approval of the whole crew, and he ordered
them to sail up to the islands under his direction. The priest Tomas, the son of Thórarinn,
said that three times, as they sailed that night, there was nothing else to be seen but rocks
immediately ahead, so far as he could tell. It was just when they had got through the islands
that they saw that day was breaking, and after this they managed to bring their ship into a
good harbour on the island of Sanday.[43]

The seamanship of the Norsemen, however, was not always of such a high order.
From time to time cases of gross negligence and a notable lack of seamanlike
precautions are recorded in the sagas. We hear, for example, in the *Ögmundar þáttr
dytts ok Gunnars helmings* how a disastrous collision off the Norwegian coast was
caused by rank bad seamanship. Ögmund and his crew had put to sea, rather late in
the summer, and got a good wind. They carried the fine weather with them across
the ocean, and towards the end of their voyage sighted land just before nightfall, the
wind blowing fresh towards the shore. The pilots on board the ship declared that it
would be more prudent to lower the sail and lay the ship to for the night, and make
the land next morning in daylight. But Ögmund was loth to lose such a good wind
and said that the moon would give them sufficient light. In the end he had his way
and their vessel sailed on through the night. But when they were but a short distance
from land they came suddenly on a fleet of longships moored in one of the sounds
between the islands; and not sighting them in time, collided with a longship and
sent her to the bottom; after which they held on their course to the harbour. Some
of the men on board Ögmund's craft grumbled that 'that was lubberly sailing'.[44]

In the *Laxdæla saga* there is another instance of bad seamanship which might have
ended disastrously but, fortunately for the crew, led to nothing worse than a
stranding.

Thorsteinn set sail before a stiff south-wester. They sailed up towards the narrows, where
the tide runs strong, and were carried into the channel known as the Kolkistu-Straum. This
has one of the strongest currents there are in the Breidafjörd waters. They presently lost
control of the vessel. What had most to do with it was that the tide was running out, while
the wind was shifty; the weather was showery, with stiff squalls when the showers broke
and next to no breeze in between. Thórarinn was steering and had wound the steering lines

about his shoulders, because the boat was crowded and there was little room. The boat was loaded with household goods for the most part, and the cargo piled up high. . . . The boat made little headway, because the current was setting violently against them. Presently they ran on the shallows without wrecking the vessel. Thorsteinn gave orders to take in the sail in a hurry.[45]

But cases such as these, indeed, were exceptional. Though it is impossible within such small compass to do justice to the consummate seamanship of the Norsemen—a race which produced some of the boldest and most resourceful mariners in the world's history—enough has been said, perhaps, to show to what a high level of skill and experience these crews had attained in the Viking age.

The ancient poetry and prose of the Norsemen, abounding as they do in seamanlike terms and phrases, lucid and terse in the highest degree, bear witness to the maritime predilections of this people. Old Norse—to a far greater extent even than modern English—is *par excellence* the language of the sea. A wide and workmanlike vocabulary covers every phase of life afloat: the stowage of cargo; the berthing of passengers; the setting of watches; messing arrangements; boat-work, watering, and provisioning; fast passages and long; the signs of the weather; the strength of the wind; the motion and appearance of the sea; the set and drift of currents and tidal streams; squalls, gales, and fogs. In the early laws of Norway and Iceland, as we have seen, evidence exists of an elementary rule of the road, and there are regulations for securing a safe load line.

The immense range and scope of the ocean navigation of the North in the Middle Ages is even today, in all likelihood, not properly appreciated outside the confines of Scandinavia. The long-continued traffic between Norway, Denmark Ireland, the Orkneys, Shetlands, and the Sudreyjar, and the Faeroes, Iceland, and Greenland, serves as a complete and convincing refutation of the old fallacy—which is, nevertheless, an unconscionable time dying—namely, that it was not until the introduction of the magnetic compass in the later Middle Ages that European mariners dared to venture out of sight of the land.

The *regularity* of this ocean navigation of the Norsemen was one of its most impressive aspects. Apart from certain periods of decline and retrogression, it went on, year after year, decade and decade, century after century. Nor must it be forgotten that there were times when the sailings that we read of in the sources represent but a fraction of those actually made: for while the ventures of chieftains and other notables are described, often in considerable detail, the voyages of less distinguished men would usually go unrecorded. The whole matter has given rise to one of the most intricate and baffling problems in the annals of the sea. The question has frequently been asked: How were the Norsemen able, with such apparent confidence and certainty, to cross and re-cross enormous stretches of open sea, at a time when not merely the quadrant and astrolabe, but even the magnetic compass, were as yet unknown?[46]

That the Norsemen were guided by no such infallible instinct as has sometimes been suggested but relied upon some sort of dead reckoning which not infrequently failed them would appear from the significant expression that occurs in several of the Icelandic sagas: *hafvilla*.[47] It would appear from various passages in the sagas that when in this state of *hafvilla* a crew lost all sense of direction and 'did not know where they were going'.[48] In the *Halldórs þáttr Snorrasonar* a crew is said to have remained in this condition for a fortnight. *Hafvilla* might overtake a crew in mid-ocean or in the vicinity of land. Cases of *hafvilla* are recorded off the Norwegian, Irish, and North American coasts, as well as far out in the Atlantic. The plight of Örlyg and his men, as related in the *Landnámabók*, is perhaps the earliest known instance of *hafvilla*.[50]

The allusions in the sagas to *hafvilla* are frequently preceded by the ominous phrase, *tók af byri*, 'the fair wind failed'. Thus in the *Finnboga saga* it is stated that 'tekr af byri, ok gerir á fyrir þeim hafvillur', 'the fair wind failed and they wholly lost their reckoning'.[51] It is to be emphasized once again that the *hafskip*, or ocean-going merchantman, was essentially a sailing ship. Such vessels were wholly dependent on the wind. They might have to wait, sometimes for days and weeks even, for a good 'slant'. It followed that from the moment the wind ceased to favour them crews were in danger of losing their reckoning and of falling into this state of *hafvilla*.

In the *Laxdœla saga* there is a brief but lucid account of the conditions leading up to *hafvilla*. 'They met with bad weather that summer. There was much fog, and the winds were light and unfavourable, what there were of them. They drifted far and wide on the high sea. Most of those on board completely lost their reckoning.'[52] Sometimes a craft would be forced off her course by head-winds and gales. In the *Grœnlendinga saga* Bjarni Herjólfsson and his crew were driven southward by north winds.[53] In the *Njáls saga* they got 'bad weather', and on another occasion a ship was forced southward by 'strong north winds'.[54] In the *Illuga saga* a craft was driven westwards into the ocean by heavy weather, and the end of it was *fá þeir hafvillur*.[55] A common concomitant of *hafvilla* was *þoka* or *myrkr* (fog). In the *Grœnlendinga saga* Bjarni Herjólfsson and his crew encountered 'north winds and fogs, and they did not know where they were going'.[56] Similarly, in the *Laxdœla saga* mention is made of 'dense fog',[57] in the *Halldórs þáttr Snorrasonar* and *Þorsteins þáttr Bœjarmagns* of *myrkr oc hafvillur*,[58] and in the *Hálfdanar saga Brönufóstra* of *þokur ok hafvillur*.[59] In the *Njáls saga* we hear that they were 'overtaken by a thick fog' and that 'there was a dense fog'.[60] The *Illuga saga* mentions 'a thick fog, which did not lift for many days', following which *fá þeir veðr hörð ok hafvillu*.[61] It would seem that in all these cases the skilful dead reckoning which was a prime factor in the navigation of the Norsemen had completely broken down. They had no idea where they were or even which way they were heading.

Nor, again, was it a case of 'hit or miss' navigation. The facts speak for themselves. Throughout a period of several centuries this ocean traffic was carried on with no more than a reasonable proportion of losses over what was even in the spring and summer months one of the roughest and most perilous sea-routes in the world. For here was no region of steady trades or seasonal monsoons. In the unsettled conditions of these high latitudes the weather could never be depended upon; sudden changes of wind were of common occurrence; the heavens were often obscured behind mist or cloud; and added to all these were the hazards of cross-current navigation. The Norsemen aimed to arrive at their proper destination: the Faeroe Islands, one of the havens of Iceland, Hvarf in Greenland, or wherever it might be. If they made their landfall too far to the northward they said so. If they fetched up in Greenland instead of making an Iceland port, they said so too. Many such instances are recorded in the sagas. Bjarni Herjólfsson, sailing from Iceland to join his father in Greenland, was forced off his course and arrived off the eastern seaboard of North America; yet within not many weeks he had safely traversed a part of the ocean totally unknown to him and to his crew, and joined his father in Herjólfsnes in the new colony. It says much for the high standard of seamanship and navigation among Bjarni's contemporaries that such a feat was taken so much for granted that, instead of being acclaimed for what he had achieved, Bjarni on his arrival in Greenland found himself somewhat severely criticized for not having followed up the new discovery.

The Norsemen of the Viking age and their successors are to be accounted some of the most daring and accomplished seamen known to history. For centuries they had no compass, but shaped a course across the open sea by means of azimuths of the heavenly bodies.

By day the Northern seaman steered by the sun. Before the advent of the magnetic compass, it was, indeed, his only guide when out of sight of the land. In the zone of the midnight sun he could only distinguish his airts when the sun crossed the meridian: that is to say, either at noon, *sól í fullu suðri*, or else at midnight. Though in the sagas no mention is made of the method by which the Norsemen were accustomed to get their bearings (*deila ættir*) from the sun, it would appear from the passage in the *Grænlendinga saga* quoted above that the 'division of the horizon' was regarded more or less as a routine operation. In short, in the high North, during the summer months when such voyages were customarily made, the sun alone could guide the mariner across the waste of waters.[52]

By night the mariner steered by the Pole-star. The supreme importance to the seaman of this, the *leiðarstjarna*, or 'guiding star' is plainly revealed in a number of Icelandic sagas and elsewhere.[63]

The sounding-lead was certainly used in the North. The *Konungs Skuggsjá* refers to the 'deepest part of the ocean' between Iceland and Greenland, and in the thirteenth-century *Historia Norvegiae* there is an allusion to one of the early voyagers to Iceland 'probing the waves with the lead'.[64] The sounding-lead cannot, of course, have been so important in the navigation of the Iceland and Greenland Seas as it was in that of comparatively shallow waters like the North Sea and the Baltic. It would probably be used in the vicinity of the Faeroe banks, or when approaching the Iceland coast.

Whether or not the Norsemen possessed any other instruments or devices to aid them in their ocean navigation is uncertain, and likely to remain so. To be sure, the possibility cannot be altogether ruled out: but nothing like conclusive evidence has ever been adduced in support of such a claim.[65]

To supply an adequate explanation of these early ocean passages is not easy. The problem is mainly one of paucity of material. Though the sagas and other sources are full of information on the subject of shipbuilding and seamanship, there is little enough to be gleaned about Norse methods of navigation.

All the available evidence goes to show that the ocean navigation of the Norsemen may be resolved into three main elements. As may be seen from these sources, it was based on a very careful dead reckoning, which was occasionally checked by a crude observation of the heavenly bodies, and also by such adventitious aids to navigation as seabirds, whales, and ice-floes.

In the sagas and various other sources, the distances between a number of points in the North Atlantic are reckoned in terms of 'day's sailing', *dægr-sigling*. A word of explanation is required here. Strictly speaking, the term *dægr* meant a period of twelve hours: since, according to the *Rímbegla*, an astronomical and geodetic treatise compiled in Iceland in the late thirteenth century, in a day there were two *dægra* and in a *dægr* there were twelve hours. But in point of fact, *dægr-sigling* sometimes covered twelve, and sometimes twenty-four, hours. The matter is indeed a highly complex one, abounding in inconsistencies. The numerous controversies centred around *dægr* and *dægr-sigling* are unlikely ever to be satisfactorily resolved.[66]

The unit of measurement holds good on both sides of the Atlantic. The courses recorded by Ívar Bárdarson off the west coast of Greenland are given in the same way, in terms of 'day's sailing'; so are the courses shaped by Bjarni Herjólfsson off the coast of continental America.[67]

According to the *Rímbegla*, the term *dægr-sigling* indicates a distance, not a period of time. This distance would appear to be 144 nautical miles. It is to be observed that, whenever the word *dægr* is used in the sagas in connection with the sea, it invariably stands for *dægr-sigling*.

It would appear, as one would expect, that dead reckoning was a prime factor in the ocean navigation of the North. The Norsemen seemingly arrived at tolerably accurate results over a very wide range of distances. It has been argued by Roald Morcken that, since their distance tables show remarkably correct figures, they probably used some kind of log—possibly something akin to the method of estimating speed which was later known as 'the Dutchman's log'. He furnishes numerous examples covering the Norwegian and Icelandic coasts, the west coast of Greenland, and various Atlantic ventures.[68]

From the sailing directions cited in Sturla Thórdarson's version of the *Landnámabók* and elsewhere it would appear, as one would expect, that dead reckoning was a prime factor in the ocean navigation of the North. The Norsemen may well have used some sort of rough tally system on their voyages, as did the Mounts Bay fishermen almost down to the present day. Otherwise such calculations had, of course, to be entirely mental, since they had no charts. Their dead reckoning would have to be supplemented and corrected by the other methods of Norse navigation which will presently be described.

As has already been said, however skilful this dead reckoning it plainly had its limitations. Thus in the *Njáls saga* there is the significant statement: 'Many lands there are . . . which we might strike with the weather we have had—the Orkneys, or Scotland, or Ireland'; and later in the same saga: 'Then they quite lost their reckoning'.[69] It is necessary to exercise caution, the more so as some years ago there was a tendency somewhat to exaggerate the part played by dead reckoning in the ocean voyages of the past.[70] What may be done successfully in coastal waters is quite a different matter in a passage of over 1,000 nautical miles, far out of sight of the land, and in face of gales, fogs, and unknown surface currents. Under such conditions D.R. may become indeed 'a blind and stupid pilot'.[71] The fact is, granted a considerable element of luck, dead reckoning, and dead reckoning alone, might answer for an occasional successful passage, but never for the regular ocean navigation of the Viking era.

Another prime factor in the ocean navigation of the North was a crude and primitive form of nautical astronomy. It is only in fairly recent times or so that proper attention has been paid to this important factor; without which, indeed, it is impossible to account for the success and continuance of these voyages.

The Scandinavian colonists of Iceland and Greenland were accustomed to take note of the altitude both of the Pole-star and of the midday sun; and though most of the observations that we read of in the sagas and elsewhere were evidently made on land, there do exist certain references, as will later appear, to observations made at sea. Unfortunately, neither on land nor at sea is it known exactly *how* they made these observations. They would not, of course, be able to compute their latitude in terms of degrees and minutes; but it seems that they could work out their northing or southing by roughly measuring the height of the Pole-star, or the meridian altitude of the sun. What latitude meant to these Norsemen was simply the elevation of *Polaris*, or the midday height of the sun; the measured length of a shadow, or the azimuth of the sun at sunrise and sunset. In any case, whatever may have been the method or methods they used, it would appear from the results of their observations that they were at any rate sufficiently accurate to serve their purpose. In our own days of ultra-scientific navigation it is sometimes hard to realize that latitude can even be calculated (though, needless to say, only very approximately) by simple ocular observation.[72] Columbus on his return passage in 1493 judged that he was somewhere near the latitude of Cape St. Vincent by the height of the Pole-star. Nor was he more than a couple of degrees or so out in his reckoning.[73]

From the tenth century onwards there is ample evidence of the Icelanders' knowledge of practical astronomy. There was Thorsteinn Surtr 'who discovered the summer eke', and to whom, about 960, the revision of the Icelandic calendar was due. Shortly after the turn of the century, at Leif Eiríksson's encampment in Vinland, the approximate difference between the latitude of the newly discovered country and that of Iceland and Greenland was noted by observing the length of the midsummer day. In the early part of the eleventh century there was Oddi Helgason, who was accustomed to study the stars, and who, at any rate for a period, followed the sea. Stjörnu Oddi, or 'Star Oddi', as he is generally called, worked out a table of the sun's azimuth and made some surprisingly accurate observations of the sun's declination from the winter to the summer solstice. It is to be noted that Stjörnu Oddi's observations of the sun's altitude are expressed, not in degrees, but in *halft hvéla* ('half-wheel', or half the sun's diameter); moreover, Oddi uses the old Norse names for the airts, for example *landsuðr*, not 'south-east' : *utnorðr*, not 'north-west'. Mention must also be made of Einar Eyjólfsson in the *Ljósvetninga saga* who 'used often to walk out of nights and look at the stars and moon, and he had a good understanding of them'; and of Nicholas, abbot of Tverra, who during his travels in the Holy Land (c. 1150) discovered the latitude of a place on the banks of the Jordan by a homely, but tolerably effective, rule-of-thumb method. The use of the quadrant for measuring the height of the Pole-star (on land) is mentioned in the fourteenth century *Rímbegla*. There were also methods of observing the approximate height of the sun. We hear of expressions like *til þess sól er skapthá*, 'the sun is shaft-high above the horizon', and *lágr veggr undir sólina*, 'a low wall under the sun'.[74]

Admittedly, none of the instances referred to above is concerned with the sea and seafaring. But it is most improbable that these Norsemen, knowing as they certainly did the significance of latitude on land, would not be similarly aware of the practical importance of latitude at sea. Certain phenomena, indeed, must almost have forced themselves on their notice.[75] Thus they could hardly fail to observe that as they coasted down the long western seaboard of Scandinavia the height of the sun at noon would gradually increase; and conversely, that as they travelled northward again it would gradually decline. Similarly, on every clear night it would be seen that the *leiðarstjarna* was sinking lower and lower as they sailed southward; and conversely, that it was rising higher and higher, as they steered northward. When later in the Viking age they struck out across the Western Ocean to the Faeroes and Iceland they must surely have observed that if they got too far south they would have longer nights, and, on the other hand, if too far north, they would have shorter nights, or no real darkness at all. And again, when they travelled from Norway to England they must have noted that certain of the circum-polar stars which do not set in the latitude of Hálogaland would sink below the horizon as their ship sailed southward down the North Sea, and would rise again on their return. Moreover, they must have known of some method for determining the approximate moment when the sun reached the meridian, *sól í fullu suðri*,[76] for in the high latitudes they would of course be unable to get their bearings from the rising and setting of the sun: and such a method should have enabled them, with sufficient accuracy, to measure the sun's height.

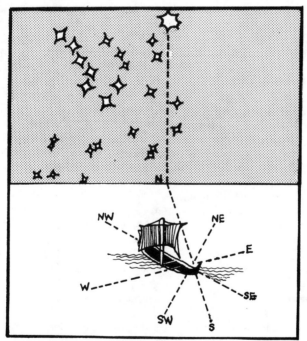

Navigation by observation of the Pole Star (Carl V. Sølver)

Consider the immense range of Scandinavian maritime enterprise in the Viking era. The mariners of that time possessed a unique knowledge of coastal navigation on both sides of the Atlantic. Unlike the seafarers of the Mediterranean, they were accustomed to make long voyages from *north to south*, as well as from east to west.[77]

From the North Cape to the Moroccan coast is more than 35 degrees of latitude; in the Scandinavian homeland itself, the Arctic and Atlantic coasts trend southward for over 12 degrees. On the famous voyage which led to the first sighting of the American continent Bjarni Herjólfsson may have traversed rather more than 20 degrees of latitude. The recorded observations of the Norsemen throughout the Middle Ages covered a wide range of latitudes: Flatey Island, off the north coast of Iceland, in c. 1140; on the banks of the Jordan in 1150; in Baffin's Bay in 1267— from lat. 32° N in the second instance to lat. 74° N in the third. Several other examples might be added to the list. Dr. Reuter was of the opinion that this remarkable knowledge of practical astronomy was in large measure fostered and developed by the far-reaching voyages of the Viking era, such as Ohthere's expedition round the North Cape to the White Sea.[78]

Earlier we have seen how as the result of contrary winds and fogs Bjarni Herjólfsson and his men had wholly lost their reckoning, so that they had no idea where they were or even which way they were heading. We have now to observe how they extricated themselves from their predicament and continued on their voyage.

As soon as the sun was seen, according to the *Grœnlendinga saga*, they were able to get their bearings, or 'divide the horizon' (*deila ættir*).[79] Now it is at least possible that in the routine operation, *deila ættir*, was included a rough calculation of the height of the midday sun. For a lengthy period they had been drifting they knew not where. But seeing how far southward by this time their craft had been driven, the sun, even to the naked eye, must have been noticeably higher in the heavens. Also the nights would be longer and darker. At all events on the following day Bjarni Herjólfsson altered course to the northward, and during the next few weeks they worked back to their proper northing and arrived eventually at their destination, Herjólfsnes in southern Greenland.

The dependence of the Norse navigator upon the celestial bodies is implied in many other passages in northern literature. Thus one repeatedly reads, as in the *Illuga saga*, how when the heavens were obscured, a crew 'had no idea in which direction they were steering'.[80] In this connection it is to be observed that, like the midday sun, the Pole-star would not only give the mariner his bearings, but might also, at twilight, by its elevation above the horizon, give him some inkling of his northing or southing, and thus assist him to keep on his proper course.[81] It may well have been for this reason that so many of the voyages of the Viking age were made in the spring or autumn, when the stars were still visible in the high North. Freydis, for example, on the return voyage from Vinland (c. 1016), loaded her ship in the early spring, made a prosperous voyage, and arrived in the Eiríksfjörd in the early summer.

In the thirteenth century the whole matter becomes clearer. The first sailing directions begin to appear. In Sturla's recension of the *Landnámabók* it is laid down that the course for Hvarf in Greenland is 'due west from Bergen'. From this it is apparent that the Norsemen knew that Bergen and Hvarf lay in approximately the same latitude. (Bergen lies in lat. 60° 4′ N, Cape Farewell in lat. 59° 45′ N.)[82]

In Hauk Erlendsson's recension of the *Landnámabók* there is an important interpolation in the sailing directions. The course, as has already been said, is now given in greater, and, so far as can be verified, in accurate and seamanlike detail. 'From Hernar in Norway sail due west for Hvarf in Greenland; and then will you sail north of Shetland so that you can just sight it in clear weather; but south of the Faeroe Islands, so that the sea appears half-way up the mountain-slopes; but steer south of Iceland so that you may have birds and whales therefrom.'[83]

Now the annual trading-ship for Greenland sailed, as we know, from Bergen. It appears, therefore, that she sailed up the coast to make her northing. On arriving off Hernar she was on the parallel of her destination, on a course that would take her north of the Shetlands and south of the Faeroes, and then a good way to the south of Iceland (as laid down in the sailing directions) to a point on the east Greenland coast about 80 miles north of Cape Farewell. To hold that course she must maintain an approximately constant latitude all the way across the ocean; and if inadvertently she were forced off her course, she must somehow or other work back to it again. For it was scarcely possible for a ship to hold her course across 1,000 miles and more of open sea, in spite of mists and gales and surface currents (not to mention variation), by means of the magnetic compass, and the magnetic compass alone. This certainly looks like latitude sailing.[84]

It can hardly escape the notice of the investigator that quite a number of these sea-routes used by the Norsemen lay along parallels of latitude. The course from Trondheim to the eastern ports of Iceland lies roughly along the 64th parallel, that between Cape Stad and the Faeroe Islands along the 62nd, and that from Hernar to the east Greenland coast along the 61st. There are also many passages in the sagas from which it would seem that the Norsemen sought the latitude of their destination, and then, with the help of a favouring wind, sailed along that parallel until they arrived at their objective. There is a good example of this in the *Grettis saga*. It is related therein that a certain vessel 'sailed south by Reykjanes and then south from the land' (Iceland). Her crew made a long and dangerous voyage of it. When they got down into lat. 62° or thereabouts they altered course and steered due east, and held that course until they reached the Norwegian coast, running aground at last on a small island called Haramsö, which lies in lat. 62° 36′ N. This also looks like latitude sailing.[85]

But above all it is to the *Konungs Skuggsjá* that we must turn if we are fully to comprehend the supreme importance to the seafaring trader of a knowledge of practical astronomy. In passage after passage throughout that work particular emphasis is laid on the value of such knowledge. It is clear that the careful study of the heavens ranks high among the 'subjects which', as the anonymous author remarks, 'especially touch the welfare of seafaring men'.[86]

In the same century (*c.* 1267) there is an interesting case of a party of Greenlanders 'taking' the sun while on a hunting expedition up the west Greenland coast. The passage plainly shows that the Norsemen, even in the remote colony of Greenland, were able, at any rate roughly, to observe the latitude of the sun *afloat*. It is worth noticing, too, that on this occasion they 'took' the sun, not only at midday, but also at midnight. 'It was then freezing there at night, but the sun shone both night and day, and, when it was in the south, was only so high that if a man lay athwartships in a six-oared boat, the shadow of the gunnel nearest the sun fell upon his face; but at midnight it was as high as it is at home in the settlement when it is in the north-west.'[87]

This is the only certain recorded case in Scandinavian literature of a crude form of latitude-determination on shipboard. It is to be noted that the second observation in particular is of special significance; since it shows that the Norsemen were able, at sea, to 'take' the sun with accuracy comparable, to some extent, with that of their observations made on land. Attention has also to be drawn to the fact that this observation of 1267, like that of the early eleventh century in Leif Eiríksson's camp in Vinland, was made, not by scholars or clerics, but by mariners.

The third factor in the ocean navigation of the North is the one summed up in the words 'adventitious aids'. It is, of the three, by far the most difficult to assess at its proper value. It consists of the information (of widely varying degrees of reliability) to be derived from the experienced observation of natural phenomena such as the movement and distribution of different species of fauna, the prevalence of fog in certain areas (for example, the Faeroes and the Grand Banks), overfalls and ice. To the practised eye, indeed, the Western Ocean cannot have been such an entirely trackless wilderness of waters as might at first sight be supposed.

This third factor must have played an exceedingly important role in transoceanic navigation. It has already been explained how the benefit of having Iceland, as it were, as a convenient 'half-way house' on the long venture between Norway and Greenland was unhesitatingly sacrificed for the superior advantages of the direct route, which was decidedly shorter and free from rocks and other dangers all the way over between the Faeroe Islands and the East Greenland coast. On his approach to the Faeroes the mariner would receive a certain amount of warning from the swell over the banks and the gathering flocks of seabirds; of his proximity to Greenland, from the significant change in the colour and temperature of the water (the Polar current), and the presence of ice.

The part played by the experienced observation of birds in ocean navigation, not only in the Viking age, but also in the time of the great discoveries, is of considerable interest and importance. The difference in the distribution of certain species, such as the fulmar and the Brunnich guillemot, might give the navigator some inkling of his northing or southing. It has been remarked that a gradually increasing proportion of the dark variety of fulmar, and also of the Brunnich guillemot, is to be met with as one reaches the higher latitudes.[88] The great skua, or bonxie, which is a very conspicuous bird, gets most of its living during the breeding-season by chasing kittiwakes and obliging them to disgorge their food, which the bonxie then swallows. Vessels bound for Reykjavik from Glasgow or Leith frequently sight the bonxie, in chase of kittiwakes and other gulls, when roughly half-way between Scotland and Iceland. When these Iceland bonxies have managed to secure sufficient food they fly home to their nesting sands along the edge of the Vatnajökull. The kittiwakes, which have many colonies on the south coast of Iceland, may also have been of service to the mariner. In certain seasons the flight-line of wild geese and other migratory species may have been of assistance. Nearer land there would be the small auks: puffins, common guillemots, etc. These, in the vicinity of island-groups such as the Faeroes which were frequently invisible through mist, must sometimes have served as a useful 'pointer'. Within the offshore zone, which extends roughly to the 100-fathom line, the typical birds are the gannet, the Brunnich guillemot in northern areas and the common guillemot in southern, the razorbill, and certain gulls, such as the lesser and great blackback. During the breeding-season, which lasts from May till August, the offshore birds pass continually from their feeding grounds in the zone to their nests, also they frequently visit their nesting cliffs at other times during the spring and summer. Fishermen of the Shetlands and Iceland are accustomed to use the guillemots as a guide when uncertain of their position.

Gannets frequently serve to pilot British trawlers to the entrance of the Pentland Firth in a fog. Mounts Bay fishermen and West of Ireland curachmen also use the local seabirds as a guide. The Vestmannaeyjar—in the medieval era one of the favourite landfalls in making Iceland from the south—are probably the largest bird station on the south coast of Iceland.

From the evidence of the sagas it would appear that the Norsemen carefully observed the birds they sighted at sea. In the *Færeyinga saga* a certain crew 'put to sea when they got a wind and went on till they saw birds from the islands'. In the *Eiríks saga rauða* Thorsteinn Eiríksson and his companions were driven eastward across the ocean till they sighted 'birds from Ireland'. In the *Biskupa sögur* a ship's crew 'drifted southward across the ocean, so that they had birds from Ireland'.[89] Similarly, in Hauk Erlendsson's sailing directions the mariner was instructed to steer 'south of Iceland so that you have birds and whales therefrom'.[90] A richer admixture of plankton in the sea to the south of Iceland, through the action of surface currents, might account for a relative abundance of seabirds and whales. As has already been said, it is related by modern Icelandic seamen that one first gets 'birds from Iceland' at a point some 150 miles south of the island, which is approximately the distance at which the course, given in the *Hauksbók*, passes by the south Iceland coast.[91]

In the Icelandic sources there are various references to sea-marks, particularly in respect of islands.[92] In the passage from the *Hauksbók* already mentioned the mariner is directed to steer south of the Faeroe Islands 'so that the sea is half-way up the mountain-slopes'. In the *Óláfs saga helga* there is a reference to a vessel steering down the west coast of Norway on a course 'so as to have the sea half-way up the mountain-sides'.[93] Similarly, vessels often made their departure from, and shaped a course for, well-known promontories, for example, Cape Stad in Norway,[94] the steep headland of Reykjanes in Iceland, the 'high land of Hvarf' in Greenland.

The colour and run of the sea would sometimes give the mariner an approximate idea of his position.[95] On two separate occasions in the *Njáls saga* crews which had been in the state of *hafvilla* were warned of the proximity of land by a tell-tale groundsea. On the second of these occasions—off Hrossey (Rousay) in the Orkney Islands—it is recorded that there were heavy overfalls, *áföll stór*, in the vicinity. 'Flosi then said that they must be somewhere near land and that that must be a shoal over which the seas were breaking.' It was thick and blowing weather. Their ship took the ground and broke up.[96] This race was apparently Rull Röst in Westray Sound. It would seem that the Norsemen were closely observant of such phenomena.

On the far side of the ocean the Norsemen expected to encounter ice, icebergs, and ice-floes. Towards the end of the long voyage to Hvarf the reflection above the horizon of the great Greenland ice-cap (the 'ice-blink') would serve as a welcome sign to the weary mariner that he would soon be in sight of land. From the thirteenth century onward, as one learns from the *Konungs Skuggsjá*, watch would be kept for the edge of the drift-ice off the East Greenland coast;[97] and (as in Iceland) certain glaciers, *jöklana*, would be used as marks.[98] To judge from the number and significance of various passages in the sagas and other sources one may reasonably infer that the Norse mariner made good use of every adventitious aid on the face of the waters.[99]

It was customary for crews in the Viking age to have with them a pilot with intimate knowledge of the waters across which they were to pass[100] and the contours of the coast for which they were bound: the tides, bottom, havens, off-lying dangers, and local weather conditions. In the sagas reference is often made to mariners who were said 'to know the land'. On ocean voyages some rough knowledge of the surface currents and the seabirds, whales,[101] narwhals, etc., would probably form part of the pilot's mental equipment. The importance of having a pilot or *leiðsagnarmaðr* among the crew is implicit in the statement made by Bjarni Herjólfsson before he started on his memorable voyage. He observed that their venture would be regarded as an imprudent one, since none of them had had experience of the Greenland Sea.[102] There is a revealing passage in the *Laxdœla saga* referred to above which clearly shows how some mariners possessed the art of the *leiðsagnarmaðr* in a pre-eminent degree.[103] An exact knowledge of the local tidal streams was essential for coastal navigation; since, as has already been explained, it was often impossible to sail against the tide. There were also dangerous roosts which must sometimes be given a wide berth, such as the Pentland Firth, Dynröst in the Shetlands, and Reykjanesröst off the south-west Iceland coast. Some of these pilots had such an intimate knowledge of their own particular stretch of coast that they were able after nightfall to steer a vessel up the long Norwegian fjords by the light of the stars.

Perhaps the chief lesson to be learned from the consideration of the various examples of *hafvilla* in the sagas—taken in conjunction with those far more numerous cases in which *hafvilla* is implied—is that we should do well to recognize the limitations, as well as the capabilities, of Norse navigation. During prolonged periods of thick or cloudy weather, by taking careful note of the wind and sea, mariners might hold their craft more or less to her proper course; and up to a point their dead reckoning, of course, would see them through. But sooner or later there would come a time when conditions of weather and sea were such that it was wholly impossible for them to keep their reckoning. Not only have we to reckon with the very heavy toll of losses recorded in the Icelandic annals and sagas, but we have also to bear in mind these frequent references to prolonged spells of drifting and *hafvilla*; to vessels tossed about on the high sea for most of the summer and autumn months, arriving at their destination only a little before, or just after, the winter had set in; and to ships which were unable to complete their voyage and had to make for the nearest land. On the other hand, too much must not be made of disasters and delays. The losses were heavy indeed in particular years; yet the fact remains that, over a period of several centuries on the long and hazardous ocean routes of the North, in most years the majority of these craft came safely to port.

All this admittedly runs counter to the popular belief that it was not until the introduction of the magnetic compass that mariners dared to strike out boldly across the ocean. Thus L. F. Salzman has declared that, 'As far as possible the medieval sailor kept within sight of the land and sailed from "view to view".'[104] Professor Parry tells of 'the mediaeval navigator's horror of the open sea', and states that 'at the beginning of the fifteenth century the navigator had no means of finding his position once he lost sight of land, and consequently he took care as a rule not to lose sight of the land'.[105] Rear-Admiral Harper has gone further and has roundly observed that, 'Before the introduction of the mariner's compass in the fourteenth century, the only practical means, among western nations, of navigating ships was to keep within sight of the land.'[106] These over-confident and sweeping statements can scarcely be reconciled with the known facts of Norse navigation.

Gunnbjörn Úlfsson, driven towards Greenland by a heavy gale, gets back from the skerries he had discovered off the unknown shore. Bjarni Herjólfsson, likewise blown off his course to the coast of North America, nevertheless succeeds in reaching his destination. Some years later, Thorfinn Karlsefni voyages freely between Norway, Iceland, Greenland, and the North American coast. In 1194 a crew sails far out into the Arctic Ocean to Svalbard and back again.[107] How would such voyages have been possible if the Norseman 'had no means of finding his position'?

Then there is the first transatlantic crossing from Greenland in 999. Considering that the direct voyage to Europe had never previously been attempted it is difficult to comprehend how Leif Eiríksson could have known what course to steer. Unless, as probably happened, knowing that the latitude of Bergen was approximately that of Herjólfsnes, he set out across the ocean with the intention of holding that latitude until he reached his objective.

As has already been said, for centuries the navigation of the Norsemen was based in the main on oral tradition and experience.[108] It was not, in fact, until the thirteenth and fourteenth centuries that the first sailing directions begin to appear.[109] But we must not, because of the relative scarcity of documentary evidence, underrate the value of that tradition and experience. *Si monumentum requiris, circumspice*. The vigorous, ocean-wide, centuries-long traffic of the North bears testimony to the amazing knowledge and skill of these mariners, little as we know of the secrets of their art. In which connection it is well to recall that, even within living memory, fishermen and others possessed of but small book-learning frequently had recourse to well-tried, rule-of-thumb methods handed down to them from their forbears which find no place in the navigation manuals.[110] And this holds good of the Iceland and Newfoundland fisheries, as well as of the coastwise traffic.

The navigation of the Norsemen was in all probability, like that of Christopher Columbus during the voyage of 1492–3, a skilful combination of dead reckoning and latitude-sailing. By the crude observation of celestial bodies they were able, not merely to 'distinguish the airts' (*deila ættir*) but also to determine their northing or southing with sufficient precision to keep on their proper track; and if they should be set too far to the northward or southward, through stress of weather or the action of surface currents, only by thus determining their position would they be able to recover it again. Though they could no more than guess at their longitude, they could at least be tolerably sure of their latitude. They could cling to the parallel of their destination, as it were, and follow it across the ocean till they reached their objective. This at least would seem to be the only possible explanation of their safe arrival at their journey's end, on so many occasions, after prolonged periods of drifting and *hafvilla*.

Judging by the testimony of Hauk Erlendsson in his recension of the *Landnámabók*, the lodestone must have come into general use on the northern sea-routes at some date before 1300.[111] There is no reference to it in the earlier versions of the *Landnámabók*; nor is there in the thirteenth-century Icelandic sagas, or in the contemporary *Konungs Skuggsjá*.

With the advent of the magnetic compass in northern waters, about the middle of the thirteenth century, the risk of falling into the state of *hafvilla* was of course considerably diminished. Henceforward the mariner was independent of celestial bodies for knowing at least his direction. Despite fog and overcast skies he could steer confidently across the open sea and put his trust in a device which, as experience had taught him, was a guide as sure and constant as the North Star. It is significant that from this time on no mention is made in contemporary sources, like the *Laurentius saga*, of any case of *hafvilla*.[112] This may reasonably be attributed to the introduction of the compass. At the same time, it must be emphasized, there is no reason to suppose that the Norsemen abandoned the traditional methods of ocean navigation; but rather the contrary. As had already been said, the sailing directions for the Greenland passage quoted in the fourteenth century *Hauksbók* appear to have been based on knowledge of relative latitudes rather than on compass courses.[113]

It cannot be too strongly emphasized that latitude sailing was the underlying principle of all ocean navigation down to the invention of the chronometer. Already in the Middle Ages it was practised by Arab pilots sailing to and fro across the Indian Ocean.[114] It was practised by Christopher Columbus on his voyage of 1492–3. It was practised by Vasco da Gama and his successors sailing to the East. The greatest of the Elizabethan navigators, John Davis, returning from his voyage to the North-West in 1587, sailed to the latitude of the Channel, and then shaped an easterly course until he arrived off his home port, Dartmouth. The Earl of Cumberland, bound for the Azores in 1589, first sailed down to lat. 39° N (which was the latitude of his destination), and then steered west for the Azores. Many other cases might be quoted from this and the following century.[115]

Precisely how far back the Norsemen may have practised latitude-sailing it is of course impossible to say with certainty. But judging from the range and volume of the traffic which was constantly carried on between Norway and her overseas settlements at the beginning of the eleventh century, there is no reason to suppose that the holding to an approximate latitude presented any more difficulty to the contemporaries of Leif Eiríksson and Thorfinn Karlsefni than it did in the era of *Sturlubók* and *Konungs Skuggsjá*. In particular it is difficult to escape the conclusion that the calculation of their northing or southing by comparison of the sun's height had long been familiar to the Norsemen.

16

The eclipse of Norwegian shipping

IN MANY RESPECTS the reigns of Hákon Hákonarson (1217–63) and of his son and successor Magnus Berfœtt (1263–84) mark an era in the history of the kingdom. The power and prosperity of medieval Norway had never stood higher. Her overseas possessions—the Sudreyjar, Man, the Orkneys and Shetlands, the Faeroe Islands, and, later on, Iceland and Greenland—formed an integral part of the kingdom of Norway. The most distant fjords óf Finmark, Iceland, and Greenland were linked up with her wide-ranging network of sea-routes. Commercial treaties had been concluded with England, Novgorod, Lübeck, and several other German sea-ports. To England, her principal market, she exported timber, tallow, furs, *skreið*, and herrings; from England she received in exchange chiefly grain and malt, also cloth, ale, and honey. This traffic, which was of considerable importance to both countries, continued to increase until it was ultimately brought low by the Hanseatic League in the following century. Her shipping plied to and fro in the North Sea, the Baltic, and the Atlantic Ocean. The era of Hákon Hákonarson saw development of an energetic, prosperous, and influential merchant class.

In no other country of the North did commerce play such a vital part in the national economy as in Norway, where so many essential commodities had to be imported from overseas, from both her own colonies and foreign countries. In the reign of Magnus Berfœtt the volume of Norwegian trade reached the highest peak it was ever to attain until modern times. Even later, voyages of exploration and discovery were still being made.[1]

Bergen, situated at the head of a deep commodious fjord and sheltered by lofty mountains, was at this time one of the greatest ports in western Europe. It had long supplanted Trondheim as the centre of the stockfish trade. There is a lively description of Bergen, with its harbour full of shipping and its quays piled high with stockfish and thronged with a cosmopolitan crowd of merchants and mariners, as seen by a party of Danish crusaders in the previous century.[2] Matthew Paris, who visited the place in 1248, counted upwards of two hundred ships lying in its harbour.[3] 'I see here', declared Cardinal William Sabine at the coronation of Hákon Hákonarson, 'so many men from foreign lands and such a multitude of ships that I have never seen a greater number in any harbour; and I believe that most of these ships have been laden with good things for this country.'[4]

Hákon Hákonarson did much to foster commerce and encourage the development of cities. He founded the town of Marstrand, erected a royal hall in Bergen, strengthened the coastal defences, and improved the harbour of Agdanes at the entrance of the Trondheimfjörd. He was a generous friend and patron of the Icelandic scalds and saga-writers; most of whom, at one time or another, he welcomed at his court. The *Hákons saga Hákonarsonar* presents a vivid and attractive picture of his reign.

Hákon Hákonarson was the last of his line to rule over an undiminished inheritance. Under his sway Norway still maintained her insular policy and for the most part held severely aloof from the affairs of the Continent. For four hundred years, from Harald Hárfagri to Hákon Hákonarson, Norway had been essentially a maritime power, the strongest in northern Europe. Her ships and her trade were all in all to Norway. But after Hákon's death there came a change. Recent developments had demonstrated that Norway could no longer afford the effort to retain the Hebrides; and Magnus Berfœtt sold both these islands and Man to the King of Scotland.

Under Eirík Magnusson and Hákon Magnusson the traditional policy of isolation was abandoned, and Norway plunged into the wars and intrigues of the Continent. Worse still, the latter granted favours and concessions to his country's most formidable enemy, the German Hanse, who before the middle of the thirteenth century had reduced both Norway's naval strength and the commerce which was her life-blood.

The fourteenth century was an era of progressive national decline. The promise of Hákon Hákonarson's reign was not fulfilled. Hákon Magnusson was the last male of the old royal line. On his death in 1319, though the Norwegian fleet was still the strongest in the North, the days of Norway as a maritime power were numbered. Her craftsmen still followed the old tradition of ship-construction handed down to them from the Viking age. It was not, in fact, until the middle of the century that longships were finally discarded and fighting craft of the new type constructed; but by then it was too late. Her merchant vessels also were becoming fewer and smaller. She had lost much of her centuries-old traffic with England. Her diminishing commerce and shipping were only one aspect of a general retrogression. With the extinction of her royal house the old leadership was lost. Weakness, lethargy, torpor, and stagnation were setting in on all sides; there was a significant falling-off in literature and art as well as in government and business: the whole process of decline being accelerated by the devastation wrought by the Black Death. Worst of all, the stranglehold of the German Hanse upon the national economy was steadily tightening.

The onset of the German merchants in the North had in the twelfth century been energetically resisted by Sverri, King of Norway. A century later the Germans were expressly prohibited, by the ordinance of 1294, from sailing north of Bergen. In the face of the growing demand for stockfish, however, more and more of the export trade in *skreið* fell into German hands. The fact is that Norway, as early as the middle of the thirteenth century, was becoming increasingly dependent on imported corn. The growth of the rich cod fishery of northern Norway greatly accelerated these developments; as did later the Black Death. The Nordland stood in considerable need of the grain which the Hanse sea-ports, situated as they were in convenient proximity to fertile corn-growing regions, were well able to supply; and the Hanse, in consequence of the contemporary progress in ship-design, could handle bulky cargoes on a scale never before attempted in the history of the North. The chain of causation was in fact economic and political rather than maritime.[5]

The fortunes of the German Hanse were built, first, on the fisheries, and, later, on seaborne commerce. The Hanse were essentially carriers, not manufacturers. The towns lived on, and by, the great trade-route of the North which extended from Novgorod in the east to the ports of the peninsula in the west. In the thirteenth century Bruges was the principal trading centre of northern Europe. There the Hanse had established their factory for the lucrative Flanders trade, with lesser marts in London, Boston, Lynn, Bergen, and Novgorod. At the same time the Hanse cities were drawn closer together by the ties of common interests, especially the combination of Wendish towns under the leadership of Lübeck. This group was at the head and heart of the Hanseatic confederation during the greater part of the fourteenth, fifteenth, and sixteenth centuries. It was this group which controlled the herring trade of Skania, and in Bergen gained a virtual monopoly of Norway's foreign trade.[6]

The formerly flourishing Norwegian shipping in the Baltic and North Sea was now progressively supplanted by the broad, roomy cogs of the German Hanse. After the introduction of the stern rudder and improved rigging in the thirteenth century the size of the cog steadily increased. The cog[7] was a slower sailer; but, carrying no oars, required a much smaller crew. Carvel-built, with a new type of hull constructed on quite different lines, it could load far more cargo than the kaupskip.

The evolution of the new type of vessel had, in fact, made possible a new form of trade: the transportation of bulky products like corn and fish. Bulk-transport had become essential to the conditions of life in northern Europe. The Norse hafskip, as has been said, was pre-eminent for its sailing and seaworthy qualities; but its cargo capacity was comparatively small. The economical operation of the cog improved as its size was increased. Norway saw her shipping decisively outclassed. The new economy that was arising in that land was based on the exchange of foreign corn for the dried cod, or skreið, of northern Norway and, somewhat later, of Iceland. From now on the Norwegian fisheries were systematically exploited by the enterprise and initiative of the German Hanse. It was, in fact, not so much a matter of the deliberate destruction of Norwegian trade and shipping by the Hanse, as of the inauguration and development of a kind of commerce which had scarcely existed before.[8]

Following the Treaty of Stralsund in 1370 the commercial supremacy of the Hanse was absolute and uncontested. The Norwegian merchants were gradually forced out of business. The Hanse cities had almost everything the Norwegian towns lacked—self-government, a united and consistent policy, superior organization, far greater capital resources and mercantile experience, and a more modern trading-fleet. Towards the close of the fourteenth century Lübeck's trade with Bergen was on a very large scale, being several times as large as the Norwegian trade with Lynn in former years. On the east bank of the Vaage—where throughout the medieval era the more important trade and shipping usually resorted—the Germans had their quays, which were commonly known as the German Quays. The Hanse kontor in Bergen, comprising about two dozen substantial blocks of buildings, which provided both living quarters and business premises, with its three thousand merchants, masters, and apprentices, formed, in fact, a kind of state within a state. It was here, laboriously poring over their ledgers in the dark little offices by the Vaage, that the Germans gradually gained possession of the overseas trade of Bergen, by virtue of their careful industry, skilful manipulation of credit, and genius for organization.[9]

By far the most formidable rivals of the German Hanse in this region were the English traders of the east coast ports, especially those of Lynn, who had their own guild and trading-station in Bergen. But, as has already been said, the English were gradually but surely being excluded from that town; and for the next hundred years or so the Hanse exercised what amounted to almost complete control over Norwegian commerce and shipping. The complaints of the Norwegian merchants and council were alike unavailing. The German Hanse were the undisputed masters of Bergen; they practically monopolized its foreign trade; and, in the fifteenth century, their *kontor* rose to the height of its power and prosperity.[10]

The fourteenth century, therefore, had seen the loss of most of Norway's foreign trade to the Hanse. The fifteenth century was to witness the rapid decline of her traffic with her overseas possessions, and the eclipse of her shipping for hundreds of years to come. The once vigorous trade between Norway and Iceland, which had endured since the days of the Settlement, was virtually extinguished. In 1412 the Icelandic annals sadly recorded: 'No tidings from Norway to Iceland'.[11]

III

THE OCEAN VOYAGES OF THE ENGLISH AND HANSE

17

The first English voyages to Iceland

A GOOD DEAL of obscurity surrounds the first arrival of English mariners in Iceland. According to Richard Hakluyt's *Principal Navigations, Voyages, and Discoveries of the English Nation*, the men of Blakeney fished off the coasts of Iceland in the reign of Edward III; but the statement is unsupported by any kind of evidence. It is recorded in Blomefield's *History of Norfolk* that 'Robert Bacon, a mariner of this town of Cromer, found out Iceland'. Blomefield does not say where he got this information. Professor Carus Wilson in her essay on 'The Iceland Trade' implies that mariners from our east coast ports found their way to Iceland as early as the twelfth century.[1] But this statement is not based on any contemporary authority.

What may perhaps refer to the chance arrival of Englishmen in those distant waters is an entry in the Icelandic annals under the year 1397 relating to a brawl in the Vestmannaeyjar with certain unnamed foreign merchants (*vtlenskvm kaupmonnum*) from six ships which had apparently wintered in the islands.[2] But proof of this is still to seek. The adjective is one which might apply to Germans and Hollanders as well as to Englishmen; and the riddle is unlikely ever to be answered. An earlier entry in the Icelandic annals (1337), however, relates that a crew of fifteen Scots, sailing from Bergen to Scotland, were driven off their course and arrived on the coast of Iceland; and that of these fifteen, no more than five survived. It may very well be, therefore, that in some such manner as this an English crew likewise 'found out' Iceland.

Another possible explanation is that some Norwegian or Icelandic pilot may have conducted an English vessel to the coasts of Iceland; or, again, that there may have been Englishmen on board some Norwegian or Icelandic vessel which made this voyage. After all, Lynn and several other of our east-coast ports had a long-standing connection with Bergen, whose merchants enjoyed the monopoly of commerce with Norwegian possessions overseas. Bergen, in fact, is by far the most likely point of contact in this regard. What is in any case reasonably certain is that in about 1400 the sailing directions for the voyage to Iceland were known to a number of Englishmen.[4]

Formerly the English mariner had been for the most part tied to the coast, a 'rock-dodger', steering from 'view to view', and seldom out of sight of the land. Year by year, in fair weather and in foul, sturdy coasting skippers of the type of Chaucer's Shipman had navigated their cogs, barges, or balingers around the long, indented coasts of Britain or fared forth across the narrow seas to the ports of Holland and Flanders. They acquired in these activities a familiar knowledge of the intricate coastal navigation which could only come from the accumulated experience of generations. For long this knowledge was neither written nor read: but was passed down by hand and mouth from generation to generation. In the later medieval era the range of these ventures steadily increased.

In the first decade of the fifteenth century, during the spring of 1408 or 1409, according to an English source, English fishermen had begun to work the Iceland fishery;[5] and had thereby opened a new and significant chapter in the history of English maritime enterprise. It was by far the longest ocean venture made by Englishmen until the discovery of Newfoundland.

At this stage it becomes necessary to inquire into certain political and economic factors which played a decisive part in opening up traffic between England and Iceland. In the early years of the fifteenth century various developments were fast coming to a head. At the bottom of the matter was the situation in Bergen. The Germans had not only secured the foreign trade of Norway, but had also driven out their most formidable competitors for that trade, the English merchants, especially those of Lynn. Though the English connection with Bergen was never completely broken, it suffered a great falling-off. It is significant that the *Libelle of Englyshe Polycye*, which devotes a dozen lines to the Iceland trade, does not so much as mention Norway. From now on the Hanse practically monopolized the trade of Bergen, including the all-important export of stockfish.

It was to purchase stockfish that so many east-coast vessels crossed the North Sea. But the situation in Bergen, from the standpoint of English interests, was rapidly deteriorating. The town was sacked by German pirates in 1393, and again in 1394. The position of English merchants was becoming untenable. What was only too certain, by the opening of the fifteenth century, was that the Hanse would never leave them in peace in Bergen. It accordingly became necessary to look for other sources of supply. The lure which drew English mariners across the ocean to Iceland not many years afterwards was unquestionably the 'comodius stokfysshe of Yselonde'.[6]

The stockfish sold in Bergen came partly from the northern parts of Norway (especially the Lofoten Islands) and partly from Iceland. Since 1294, as we have seen, the Staple had been at Bergen, where the royal dues were collected; and foreigners were strictly forbidden to sail to Iceland and the other dependencies of Norway. The part played by stockfish in the economy of western Europe can scarcely be exaggerated. Not only did the Church forbid the consumption of flesh on many days in the year, but fresh fish was often unobtainable in the winter. In the second half of the fourteenth century immense quantities of *skreið* were shipped from Iceland to Norway. But with the steady strangulation of Norwegian trade and shipping by the Hanse communication between Iceland and the mother country became sporadic and uncertain. For many of the amenities of civilized life Iceland was wholly dependent upon shipping from Norway. In vain the Icelanders protested that they were not receiving the six ships yearly that had been agreed upon when Iceland was united to Norway in 1263. It was, therefore, at an opportune moment that the English came to Iceland.[7]

In 1412 there occurs the first mention in the Icelandic annals of the arrival of Englishmen in the country.

A ship came from England east off Dyrholmaey. Men rowed out to them, and they were fishermen from England. That same autumn five of the Englishmen parted from their companions and came on shore east at Horn out of the boat, and said they wished to buy food, and said they had been starved in the boat for many days. These five Englishmen were here in the land that winter; for the boat was gone from them when they came back, and

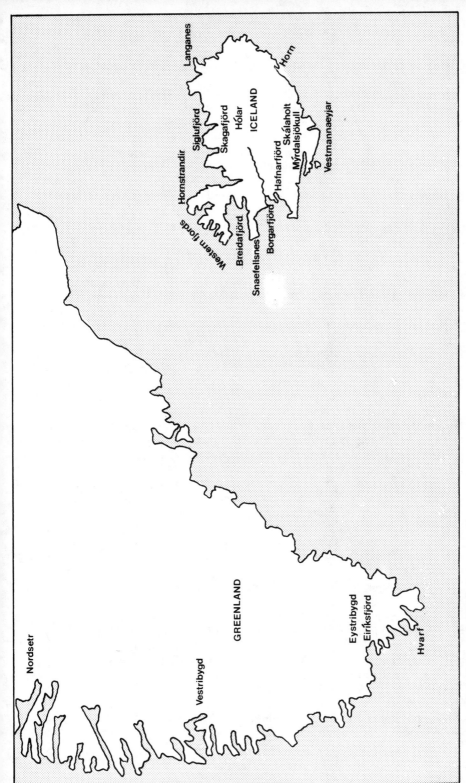

Iceland and Greenland in the later Middle Ages

also many who were out in it. One was lodged in Thyckvabœr in Ver, another at Kirkjubœr; three to the east of Orœfi. No news came from Norway to Iceland.[8]

In 1413 the fishermen were followed by traders. A merchant ship came from England to Iceland. Her master was a man called Richard, and he had with him the King of Norway's letters to the effect that he might sail with his wares to his realm without paying toll. 'He landed from his ship east at Horn, and so rode to Skalaholt and back under Eyjarfjöll. There he got on board his ship and sailed in her to Hafnarfjörd.' Many Icelanders came down to the shore to buy the wares in her; and shortly afterwards the ship sailed away, taking with her the five fishermen who had wintered in the island. That same summer, according to the Icelandic annals, there sailed out from England '30 fishing doggers or more'.[9]

It was during this year, too, that we first hear of the lawless predatory behaviour for which these mariners of the east-coast ports were later to become notorious.[10] Some of the fishermen who had sailed round to the north of the island went ashore and took some cattle, for which (not knowing the language) they 'put down money'. But another party who had landed for the same purpose on Papey, an island off the south-east coast, went off with the cattle without paying for them. In the circumstances it is not to be wondered at that the Icelanders were displeased at the appearance of foreign fishermen off their own fishing-banks, apart from these depredations: but the English merchants were welcome from the start.

A word must be said about the 'fishing doggers' referred to above. The dogger played no inconsiderable part in English maritime enterprise in the later medieval era. Originally an importation from Holland, it was a type of craft almost peculiar to the east coast; and though nothing is known about either its construction or rigging it is certain that it was both seaworthy and of a fairly substantial size, ranging from 30 and 40 tuns to sometimes 80 and even more. As a rule the dogger was used for fishing. Numerous references to these doggers are to be found in contemporary sources throughout the fourteenth and fifteenth centuries. In 1337 the Icelandic annals record that a Norwegian abbot and his men were murdered, and their corpses pitched overboard, by a gang of English 'doggerers'. In a statute of 1358 reference is made to 'all the ships called doggers and lodeships, pertaining to the Haven of Blakeney'. A dogger of Cley was impressed for the King's service in 1373. In the reign of Henry V there are further references to doggers from east-coast ports. Doggers are first heard of in the North Sea; but early in the fifteenth century, as has already been said, and for well over a century to come, large numbers of them were used in the Iceland fishery.[11] It was in all probability an east-coast dogger which had arrived off the coasts of Iceland in 1412.

It is of no small significance that the opening-up of this English traffic to Iceland coincided with the beginning of a revolutionary advance in shipbuilding and sail-plan. Approximately by the year 1400 vessels—both large and small—were being built which, not many years later, were to make the long venture to Iceland. Many technical terms familiar to us today were coming into use. 'Poops', 'forecastles', 'hatches', 'davits', 'bowsprits', 'yards', 'shrouds', 'stays', 'backstays', 'swivels', 'blocks', and 'slings' are among the terms to be found in early fifteenth-century inventories and expense accounts. But the most remarkable advance of all was in respect of ship-design and rigging. Quite early in the fifteenth century—at any rate, on the east coast—the two-masted merchantmen had begun to supplant the old, single-masted type of vessel. There is an interesting carving of such a two-master on a stall-end, formerly in St. Nicholas's chapel at Lynn, and now in the Victoria and Albert Museum. The vessel depicted there, with its finer lines and sharp, sturdy bows, is a distinct advance on the old, basin-shaped cog. Moreover, the carving shows that while the mainmast was square-rigged as of old, the mizen was fitted with a lateen, or fore and aft sail.[12] Rather later, the two-masted vessel is portrayed in the mural paintings of the Lady chapel at Winchester, and of Breage church, Cornwall; and also in the painted glass of Hillesden church.

At the same time there had been a parallel advance in the knowledge and practice of navigation. For more than a century and a half the magnet used as a compass had been known all over Europe, and had served as a valued aid to navigation in thick or cloudy weather. By about the year 1300 the compass rose was in use among Italian seamen in the Mediterranean. This was a marked improvement on the earlier and more primitive forms of the instrument, such as the needle (magnetized with a lodestone) floating on a chip or straw in a cup of water. The time of its arrival in the waters around the British Isles is uncertain. But, at any rate, a compass of thirty-two points was known to English mariners in the second half of the fourteenth century, as appears from a well-known passage in Chaucer's *Treatise on the Astrolabe* (*c.* 1390): 'Now is thyn orisonte departed in 24 parties by thy azimutz, in significacion of 24 partiez of the world; al-be-it so that shipmen rokne thilke partiez in 32.' From 1400 onwards there are numerous references to the compass, and the parts of a compass, in English and Hanse shipping.[13]

The mariner of those days had much in common with the masters of English sailing coasters in the early years of the present century. Like the Shipman in the *Canterbury Tales*, the fifteenth century skipper knew his 'herberwe and his mone'.[14] He knew from long experience that, at each haven along the coast, with the moon in a certain quarter of the heavens it was always high water. A rather more obscure—but highly important—branch of the mariner's art related to the Pole-star. This, for centuries after the introduction of the magnetic compass, continued to be of vital importance to the medieval mariner. Reference is made to it in Icelandic poetry, the *Maríu saga*, the *Gísli saga*, the *Canterbury Tales*, *Morte d'Arthur*, *Piers Plowman*, and *Knyghthode and Bataile*. It was the *sjóvar stjarna*, *flæðar stjarna*, the 'Sterre of the Sea', the 'lode sterre', the 'Schyppman sterre', *Stella Nautica*. An extensive body of lore and tradition had grown up around the constellation.

If, has been surmised, the English learned the course for Iceland from Norwegian mariners at Bergen or elsewhere, they may also have learned something of latitude sailing: namely, of first seeking the parallel of their objective, and then sailing along it until they reached their destination. This would involve too a knowledge of seasonal weather conditions and prevailing winds.

All this goes to show that the methods of navigation practised by English masters and pilots were not based on general scientific principles, but on personal experience and observation, and on the traditional, rule-of-thumb devices and expedients handed down to them from their fathers.

Thus the progress achieved in practical navigation was essentially a better understanding and more expert application of methods already in use. No evidence exists as to the use of any astrolabe or quadrant on board English vessels of the time; nor is there any reason to suppose that such instruments were ever used at all by

English mariners in the medieval era.[15] Yet all this time English vessels were venturing further and further afield: across the open sea to Gascony and the peninsula; across the broader stretches of the North Sea; out into the Atlantic to Iceland, and occasionally from Iceland to the northern parts of Norway.

Familiar with the courses, bearings, rocks, sands, and tidal streams of his own and the contiguous continental shores, the English mariner had now to face the dangerous navigation of the Icelandic coasts and fjords. He had to contend with the complicated conditions of the currents and tidal streams,[16] the hazardous roosts off certain of the promontories (for example, Reykjanesröst and Langanesröst), the blinding mists which cling to its east and south-east coasts, and the strong magnetic disturbances which, in so many places, rendered his compass an unsafe guide. It says a great deal for the relatively high standard of seamanship and navigation which now obtained among the mariners of the east-coast ports that so large a number of fishing doggers were immediately ready to make the long voyage to Iceland in 1413. There was apparently no difficulty in procuring masters for the ocean passage; whole droves of them were forthcoming in those early years. They brought with them their native hardihood and resource, their skill and experience as sailors and fishermen, their swift adaptability to the new conditions.

How great were the perils that awaited them on the 'costes colde' may be gathered from an entry in the Icelandic annals under the year 1419 (which incidentally gives also some idea of the proportions that the English traffic had by then assumed). 'Then came on Maundy Thursday such a hard gale with snow, that far and wide all round the land English ships had been wrecked, no fewer than twenty-five. All the men were lost, but the goods and splinters of the ships were cast up everywhere. The gale came on a little before breakfast, and lasted not quite to noon.'[17]

In 1414 there came more merchantmen. Five English ships arrived off the south coast of Iceland, and all ran in to the Vestmannaeyjar. 'There came out there in a letter sent by the King of England to the commons and all the best men in the land, that trading might be allowed to his men, and especially to the crew of the ship which belonged to him.' The Icelanders referred to the Staple at Bergen. But the English would have nothing to do with that; and afterwards trading began. In the same year letters arrived from Eirík, King of Denmark, forbidding all trade with foreigners with whom it was not customary to trade. Next summer no less than six English vessels lay in the Hafnarfjörd. It is stated that one of these crews stole some stockfish at Rosmhvalanes and also in the Vestmannaeyjar.[18]

In 1415 King Eirík protested to Henry V of England that, as of old, no foreigner was permitted to voyage to Iceland, the Faeroes, Shetlands, or other possessions of the Crown of Denmark, with intent to fish or to trade, under pain of death or loss of goods; and he complained of depredations committed by such intruders in the said lands.[19]

The English council, with the new French war on its hands, did not care to risk an embroilment with Denmark. In the winter of 1415 it was accordingly proclaimed in Berwick, Newcastle, Whitby, Scarborough, Hull, Grimsby, Lynn, Great Yarmouth, Gippewich, Blakeney, Dersingham, Burnham, Cromer, St. Botolph's, and Orwell that 'until the end of one year to come no subject shall repair to island parts of the realm of Denmark or Norway to fish or for other causes to the prejudice of the King thereof, otherwise than according to ancient custom, and especially to Iceland'. This prohibition of the Iceland traffic occasioned an indignant petition to Parliament. It was contended that since the fish had abandoned their former haunts, 'as is well known', English fishermen had searched in other seas and had found an abundance of fish off the coasts of Iceland, where they had fished for six or seven years past. The only outcome of this petition, however, was the traditional form of rejection, *Le Roy soy avisera*.[20]

In practice the prohibition restricted all traffic to the Staple at Bergen. But the official veto was generally ignored; and the English continued as before to venture to Iceland, where they sought the governor's permission to fish and to trade. Thus in the summer of 1419 the governor, Arnfinn Thorsteinsson, granted leave to the master and crew of a vessel called the *Christopher*, then lying in the Hafnarfjörd, to buy and sell in the Vestmannaeyjar and anywhere else they would, and also to fish anywhere they wished. The visitors gave good prices (about 50 per cent higher than the Norwegian) for Iceland's principle article of export—*skreið*, or stockfish; and the goods they brought with them were no less welcome. In 1415 the governor himself sailed to England in one of their ships, taking with him fifty lasts of stockfish, besides much coined silver. From England the Icelanders got what they needed most of all; namely, corn and cloth. For years the traffic prospered in spite of the ban. Denmark lay afar off and her King's writ ran but lamely here; while the English government did nothing in particular to make the ban effective.[21]

For their part the Icelanders made no bones about admitting that there had been traffic with foreigners, but reminded King Eirík that it had been agreed that 'six ships should come hither from Norway every year, which has not happened for a long time, a cause from which Your Grace and our poor country has suffered most grievous harm. Therefore, trusting in God's grace and your help, we have traded with foreigners who have come hither peacefully on legitimate business, but we have punished those fishermen and owners of fishing-smacks who have robbed and caused disturbance on the sea.'[22]

No mention is made in the early stages of the English traffic to Iceland of any other mariners but those belonging to our east coast. It would seem from the anonymous *Libelle of Englyshe Polycye* (c. 1436) that Bristol had no commerce with the 'costes colde' until about 1424. The doggers, which usually formed a substantial part of the English shipping bound for Iceland, were distinctively an east-coast craft. It was nearly always these men of the east coast, too, who made the trouble in Iceland, of which so much is heard in the ensuing decades.

For more than sixty years the bulk of the Iceland trade was to remain in English hands, while our fishermen practically ousted the Icelanders from their own fisheries. The traffic was a nursery of hardy seamen. The long ocean venture and the intricate coastal navigation tested their skill to the utmost. Their vessels, too, had to be strongly built and well rigged to stand the buffeting of Atlantic seas and winds. The importance of the Iceland sailings, as a factor in English maritime enterprise, can scarcely be set too high: they were the first real deep-sea ventures in our history. The Iceland traffic may accordingly be regarded as marking the half-way stage between the purely coastal navigation of Chaucer's Shipman and the transoceanic ventures of certain unknown Bristol mariners in the second half of the fifteenth century.

18

The English trade with Iceland

BY THE SECOND quarter of the fifteenth century it was not only our east-coast ports that were trafficking with Iceland. From about the year 1424 Bristol had similarly begun to send ships to that island, as may be seen from the well-known lines in the *Libelle of Englyshe Polycye*:

> Of Yseland to wryte is lytill nede
> Save of stokfische; yit for sothe in dede
> Out of Bristow and costis many one
> Men have practised by nedle and by stone
> Thiderwardes wythine a lytel whylle,
> Wythine xij yeres, and wythoute parille,
> Gone and comen, as men were wonte of olde
> Of Scarborowgh, unto the costes colde.[1]

It would appear from the Treaty Rolls and other sources that Bristol in the ensuing years became one of the chief centres of the Iceland trade. A steady stream of licences bore witness to the development of the regular, as did the frequently recurring complaints against the men of 'Bristow' in both Danish and English national archives, of the irregular, traffic. Though the Bristol customs accounts are for many of these years rather too fragmentary to allow of a precise estimate being arrived at, it is certain that the total volume of trade must have been considerable, even if it did not compare, in point of economic importance, with Bristol's trade to Ireland, Gascony, or the Iberian peninsula. What is highly significant, too, is the substantial tunnage of some of the ships engaged in the trade and the value of the cargoes which they brought home to England. Bristol possessed plenty of ships from 100 to 160 tuns; they sometimes carried *skreið* and other fish to the value of several hundred pounds sterling.[2]

Hull remained in the forefront of the Iceland trade throughout the greater part of the century. The names of several of her leading merchants—Robert Holm, Thomas Crathorn, Robert Stevenson, Thomas Alcock, and others—appear over and over again in the customs accounts and other records. It is certain from the value and volume of the cargoes carried that many of these craft from Hull engaged in the Iceland trade were no fishing doggers, but large merchantmen. For example, in 1465 the *Antony* of Hull brought back imports worth £536, and six years later two vessels from this port carried *skreið* valued at £642. In the reign of Edward IV the traffic became a municipal venture, conducted in the name of the Mayor and Burgesses, three or four vessels being sent out at a time.[3]

No licences, so far as is known, were ever issued to Lynn to engage in the Iceland trade. But it is apparent from the lists of cargoes in the customs accounts and the

Danish-Norwegian archives that Lynn—though scarcely so prominent as Bristol and Hull—was one of the principal centres of the traffic. Lynn remained one of the leading ports for the Iceland trade from about 1420 to towards the middle of the sixteenth century.[4]

Though detailed evidence respecting the Iceland trade of the other East Anglian ports is lacking for the early half of the fifteenth century, it is probable that already at this time considerable numbers of Norfolk and Suffolk doggers were finding their way to Iceland. At any rate various craft of Cromer and Cley are mentioned in the 1430s, and those of Blakeney in the next decade.[5]

Before long the enterprising ports of Devon and Cornwall joined in the traffic. Towards the end of the 1440s the *Marie* of Northam is recorded to have visited Iceland, and in 1447 a commission was appointed to make inquisition in Devon and Cornwall 'touching the ships, barges, balingers, and other vessels of those countries, which had sailed without leave to parts of Iselond'. About this time the *Cristofre* of Fowey made several voyages to Iceland, and the *Julian* of the same port sailed thither in 1461. Among other south-coast ports which engaged in the traffic were Southampton and Sandwich.[6]

Other regions of the British Isles were also represented in the Iceland trade. In Wales, Pembroke, Chepstow, and Swansea are mentioned in connection with the traffic. Scots as well as Englishmen are stated in the Danish–Norwegian archives to have refused to pay the Danish King's dues in Iceland. The only Irish port directly mentioned is Drogheda, but it would appear that the numbers of Irishmen who from 'havenes grete and godely bayes' set out for Iceland was not inconsiderable.[7]

No precise statistics as to the total volume of trade, or even of the sailings from any one port over a lengthy period, are available. The fact that particular ports were prominent during particular years or decades is not a reliable criterion. The customs accounts are by no means continuous; there are numerous and serious omissions; this holds true even of the Bristol customs accounts, which are rather more complete than those of either Hull or Lynn. There is, however, more than sufficient evidence to enable one to say with confidence that the passage to Iceland was regularly made, by vessels of various types and sizes, from all over the British Isles. In this way a substantial proportion of the seafaring population of these islands gradually gained experience of ocean navigation.

The fifteenth century witnessed a significant advance in both English ship-building and navigation, which may together be accounted a decisive factor in the expansion of the new ocean traffic.

The day of the two-master (fitted with either a main and mizenmast, or else a main and foremast)[8] was soon ended. When Henry VI's fleet was eventually dispersed and sold by the Council, the receipt given to William Soper in 1436 for the *Petit Jesus* and her gear mentioned a 'fokesail', a mainmast, a 'mesanmast', and a 'mesansail'. It is evident that the earliest three-masters were evolved well before the middle of the century. In the naval accounts of Henry VII's reign the *Mary of the Tower*, *Martin Garsia*, *Mary Fortune*, *Sweepstake*, and *Governor* had three masts, while the *Grace Dieu*, *Regent*, and *Sovereign* had four. The *Regent* carried topsails above her courses, and a topgallantsail above her main topsail. The *Mary Fortune* and *Sweepstake*, which were quite small ships, nevertheless had three lower masts, a main topmast, and a spritsail set under the bowsprit.[9] Such evidence as is available for the mercantile marine, covering as it does ships of various types[10] and sizes, suggests that the rig of merchant shipping was likewise undergoing a revolutionary transformation. Miss Burwash in her *English Merchant Shipping, 1460–1540* cites some significant examples from the Yarmouth Court Rolls. At least one of the ships concerned (1460–1) is recorded to have had two 'fuksailles' and one 'myson'. Another, ten years later, carried a 'forecastell maste' which had both a 'cors' and a 'bonette'. The *Anthony* of Hull, and two vessels belonging to Sir John Howard, the *George* and the *Edward*, all had two, possibly three, masts. It is reasonable to suppose that two- and three-masted merchant vessels were fairly common in the second half of the century.[11] In an illuminated manuscript belonging to the Hastings family, which is believed to date back to the middle of the fifteenth century, may be seen an early representation of three-masted vessels. These craft all carry a square foresail and mainsail and a lateen mizen, and some of them a main topsail; one of them also carries a fore topsail, and all set a spritsail under the bowsprit. The hulls carry long overhanging forecastles and aftercastles.[12] As Sir Alan Moore has commented, 'The great interest of these pictures lies in the advanced development that they show. At the beginning of the fifteenth century few ships seem to have had more than one mast, and yet here in a manuscript of about 1450 or so we find at least one ship as rigged almost after the fashion of a typical sixteenth-century merchantman.'[13] Later in the century, in a manuscript 'Life' of Richard, Earl of Warwick, are illustrations of both three- and four-masted vessels.[14]

In short, in the matter of detail, this development of an improved sail-plan and rigging is exceedingly obscure. But the general outline was certainly as follows. The introduction of the two-masted vessel, very early in the century, was followed soon after by that of the three-master: the mainmast carried a topsail as well as a course: and some time later the foremast, too, was fitted with a top and a topsail. For working to windward, a spritsail was spread under the bowsprit. Thus the medieval cog—bowl-shaped, with its single square-sail—had developed into a two- and three-masted vessel, square-rigged except for its lateen mizen, with greatly improved lines, and sturdy, high bows capable of standing up to Atlantic seas, altogether an immense improvement on the cog, able to sail nearer to the wind, and fitted for longer voyages, such as those to the Levant and Iceland.

The introduction of a second and third mast made possible a substantial increase in the size of ships. A good index to the general increase in tunnage[15] during the fifteenth century is provided by the statistics of the Bristol–Bordeaux traffic. Professor Carus Wilson, from a careful study of the customs accounts, has shown that between the years 1400 and 1450 the average cargo carried on that passage had almost doubled.[16] Miss Burwash, surveying the whole range of English shipping, reaches a similar conclusion.

Each of the sources examined, lists of ships hired by the crown, safe-conducts from the Treaty Rolls, the accounts of the Bordeaux wine customs, and the censuses of shipping, even the accounts of the Øresund toll, suggests that English merchant ships increased very considerably in size during the fifteenth century. Ships of 200 tuns and over had been comparatively rare at the beginning of the period; by 1520, or even by 1450, they were fairly numerous and England could even boast of some vessels of 450 tuns burden.[17]

Such statistics as are available concerning the Iceland and Finmark sailings tell a similar story. As the fifteenth century advanced, a number of quite large vessels are

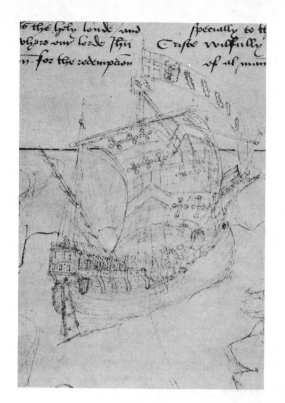

to the holy londe and ... specially to t̄...
...her ȝ our lord Jh̄u Criste wilfully
...n for the redempcion ... of al man...

*Three-masted and four-masted
vessels, from the
Beauchamp Pageant*

mentioned in documents relating to the Iceland trade. In 1439 a licence to trade to Iceland was granted to the *Katherine* of Bristol (100 tuns). In 1440 a similar licence was issued to the *Mary Redcliff* (140 tuns), and in 1443 to the *Christopher* (160 tuns), of the same port. In 1461 the *Julian* of Fowey (160 tuns) was licensed to sail to Iceland, and in 1466 the *Anthony* of Bristol (300 tuns). In 1459 a licence for six years was granted to John Gaussem and George Roche to send one ship of 400 tuns or two ships of the same tunnage to Iceland and Finmark, and in 1461 a licence for one year was granted to Thomas Napton of Coventry and John Hawkes of Bristol to send 'two ships of the realm of England of 400 tuns or under' to the same parts of Iceland and Finmark. In 1470 a licence was granted to Sir Edmund Hungerford and John Forster of Bristol to send one ship of 300 tuns or under to Iceland.[18]

Early in the 1450s the *Anthony* of Hull, variously described as of 400 and of 450 tuns, was engaged in the Iceland trade contrary to the statute of 1432; and in 1454 the Council sent orders for her arrest. For a time she seems to have escaped; for in 1456 a craft described as *l'Anthoine de Huolf*, of 400 tuns, was involved in a lawsuit at Bordeaux. In the *Registres des grands jours de Bordeaux* some important information is forthcoming concerning the *Anthony's* construction, rigging, and gear. It is evident from this source that the *Anthony* was fitted with a mizenmast, and quite probable that she also carried a foremast; for after the word 'misen' there follows a passage which is unfortunately indecipherable, and which may very well include a reference to a third mast. It is worth noticing that the *Anthony* of Hull is the only known instance of a two- or three-master employed on the ocean trade-route to Iceland.[19]

The marked progress in shipbuilding during this period is also revealed in the increasing number of voyages made in early spring and winter, not only to Gascony and the peninsula, but also across the ocean to Iceland. As early as 1404 a number of Bristol vessels set out for Gascony in mid-winter. The *Margaret* sailed on 8 January of that year, the *Katherine* on the fifteenth, and the *Welfare* and the *Cog John* on the nineteenth. The *Ive* of Bristol cleared for Iceland on 17 February 1479, and the following year the *Leonard* and the *Christopher*, also of Bristol, on the twelfth and fourteenth respectively. Seaborne trade was now being carried on on a scale unheard of during the preceding century. Ships which had sailed to Iceland early in the year would sometimes make a second voyage to Gascony, the peninsula, or the Netherlands. Thus the *Mary Foster* of Bristol, returning from Iceland in the summer of 1473, set out again for Lisbon on 16 August, and the *Ive*, on her return from Iceland on 18 July, left Bristol in the autumn for Bordeaux.[20]

In the fifteenth century, in short, more and more two- and three-masted vessels were being built. Ships were increasing in both size and number; as shipbuilding progressed larger cargoes could be carried, and longer voyages became practicable; all of which meant that an increasing proportion of English mariners were now deep-water sailors, rather than coasters. At the same time the fifteenth century saw the emergence of a specialized class of shipowners, each possessing a small fleet of ships of his own, and drawing his profits from the carrying trade. Wealthy shipowners like William Canynges, five times mayor of Bristol—'William the Great', as he was called, who is recorded to have had as many as 800 men employed in his vessels for eight years—were the predecessors of the great Tudor merchants of the following century.

During the fifteenth century the mariner's compass appears to have been generally used on the main trade-routes and in all kinds of craft.[21] It was by all accounts a prime factor in the development of the English traffic to Iceland, as may be seen from the passage previously quoted in the anonymous *Libelle of Englyshe Polycye*.

> Out of Bristow and costis many one
> Men have practised by nedle and by stone . . .

The significance of this allusion to the use of the mariner's compass on the Iceland run is that it not only shows that the instrument was in common use on this, the first of the ocean trade-routes to be followed by English seamen, but further suggests that it was carried in comparatively small vessels as well as in large—in view of the fact that from 1413 on a considerable proportion of the Iceland-bound fleet consisted of quite small fishing doggers.[22]

Miss Carus Wilson, following in the steps of Dr. Salzman, has declared that 'The familiar reference in the *Libelle of Englyshe Polycye* . . . to the needle and stone by which men found their way to Iceland can indicate only the primitive instrument, and shows that even this was resorted to only for voyages across the depths of the open ocean.'[23] The latter part of this statement is difficult to sustain in view of the many allusions to the magnetic compass on a range of sea-routes in English, German, and Scandinavian vessels throughout the fifteenth century. Nor is there really any reason to suppose that the phrase 'by nedle and by stone' refers to the needle-on-a-straw device of an earlier epoch. Lodestones are mentioned often enough in contemporary documents. The navigational instruments of the *Margaret Cely*, a vessel of some 80 tuns 'in the yere of our Lord 1485' comprised 'a grett compas, a lodston, . . . a glasse, a compas'.[24] Miss Burwash in her *English Merchant Shipping* cites a similar and earlier example.[25] Early in the sixteenth century there is another reference in *Cock Lourell's Boat*: 'One kept ye compas and watched ye our glasse. Some ye lodysshestone did seke some ye bote dyd swepe.'[26] The truth is that needles at the time were only weakly magnetized, they tended to lose their magnetism (for example, during a thunderstorm), and it was accordingly the custom to carry a lodestone—not only in the medieval period, but for centuries afterwards. The procedure has been succinctly described by Professor Morison. 'Whenever this needle showed any disinclination to "seek the north" it was remagnetized with a bit of lodestone that the captain guarded as his life.'[27] It was not, in fact, the primitive instrument, but the fully developed mariner's compass, that served to guide the shipping of Bristol and the east-coast ports across the ocean to Iceland.

Another important factor in fifteenth-century navigation was the sounding-lead and line. This was in general use in shallow seas for discovering the depth and nature of the bottom. The early history of the lead and line is very obscure. It was only in the later Middle Ages that references to its use in the seas around the British Isles begin to appear.[28] It is to be observed that the use of the lead and line was taken very much for granted in the later medieval era. As an aid to navigation it aroused none of the interested comment associated with the advent of the magnetic compass. The wonderful skill of the North Sea fisherman in navigating with the aid of this device may be traced back to this period.

Though a good deal about the mariner's compass and sounding-lead may be learned from private memorandums and ships' inventories of the day, by far the

til ye come in to iiij. fadim deep and yf it be strenp
grounde it is betwene huselfant and calle in the entre
of the chanel of flaundres and soo goo youre cours
til ye stane sixti fadim deep. than goo est northe est
a longe the see. + c.

Casting the lead

most significant and conclusive evidence is that supplied by contemporary sailing directions. In the mid-fifteenth century the first sailing directions, or rutter, in the English language appeared. There is also a certain amount of navigational information in the *Itinerarium* of William of Worcester. From these and other sources it will be seen that compass courses are given for every coast in western Europe from the Kattegat to the Straits of Gibraltar.[29]

The English rutter was copied down in the neat, clear hand of one William Ebesham, a professional scribe, and may possibly have belonged to Sir John Paston; the date of the original is unknown. From it we learn that in addition to the mariner's compass the navigational instruments then in use were the lead and line, as has already been explained, and also the hour-glass, for timing the length of the tacks. Of particular interest are the two direct courses across the Bay of Biscay from Finisterre to the entrances of the Bristol and English Channels, and also from Finisterre to Cape Clear and the Isles of Scilly,[30] showing as they do how the compass was used in conjunction with the sounding-lead. The direct navigation across the Bay to Gascony, Santiago, and other Peninsular ports—in the same way as the English trade to Iceland—was presumably made possible by the advent of the mariner's compass. There is no record of these particular passages in the pre-compass era. This old manuscript, with its recital of familiar place-names and marks and graphic descriptions, 'At the Londis ende lieth Raynoldis stone . . . Lesarde is grete stone as it were benys [beans] and it is raggyd stoon . . . Opyn of Dudman in xl fadome there is rede sande and whit shellis and small blak stonys . . . And the name of the Rok is called the Kep, and he lieth undir the watir but it brekith upon hym And the breche shewith'—is in fact the prototype of a long succession of sailing directions for the waters around the British Isles whose latest representative is the current Admiralty *Pilot*.

The course for Iceland is not given in this rutter. It is always possible, of course, that the original sailing directions may date back to a period prior to the opening-up of the English traffic to Iceland. It is on record that about the middle of the fifteenth century a vessel from Southampton sailed to Iceland by way of the Isles of Scilly and the south of Ireland.[31] From Bristol, vessels would apparently round the south and west coasts of Ireland. It would appear from sixteenth-century practice that the east-coast contingent was accustomed to sail to Iceland either by way of the Pentland Firth or else between the Orkneys and Shetlands. They were accustomed to make the return passage in good time: the danger of passing through the Pentland Firth late in the season was well understood by English mariners. Occasionally a craft from the western ports would sail to Finmark as well as to Iceland and therefore return by the North Sea route.

During the period 1430–60 the sailings to Iceland greatly increased, warnings and proclamations notwithstanding. Repeated attempts to enforce the veto were totally unavailing. In 1429 at the Guildhall in Lynn, after a stormy meeting, the council finally forbade the Iceland voyage. For a period the sailings from Lynn fell off in consequence. They did not, however, entirely cease, as may be seen from the customs accounts, and also from the numerous complaints laid at the door of the men of Lynn in certain Danish documents. Together with the mariners of Newcastle they were responsible for carrying off a large number of boys and children from Iceland. Several such cases were reported in 1429. Sometimes orders were given for these unfortunates to be taken back to their own country. It is significant, too, that there were occasional instances of Icelanders, domiciled in England, taking the oath of allegiance.[32]

In response to the King of Denmark's ordinance of 1429 the English government in the following year brought in an Act of Parliament to prohibit the Iceland voyage. Once again the proclamation was made throughout the ports. The Mayor of Hull travelled up to London to protest against the measure, and a complaint was presented in Parliament. 'Le quel Ordinance', it was said, 'est trop grevous et myschevous'; and emphasis was laid on the fact that the King's lieges and subjects 'devaunt le did Ordinance, avoient frank entrer et issir ove lour Biens et Merchandizes en Islond et diversez autres lieus deinz les Signeur(ie)s, et Territories suisditz, et loure frank pessheir et merchandiser as parties suisditz, al graunte profit a toutz les Lieges et Sugettis nostre Seigneur le Roy . . .' As for the Staple at Bergen the objection was raised : 'illeoques sount pris et detenus saunz cause resonable, a loure tres graunt anientisement et finall destruction.' The royal decision, however, was the inauspicious 'le Roy s'advisera'; and all direct traffic with Iceland was forbidden by law.[33]

Nevertheless, it was possible for merchants to obtain a licence either from the King of England or from the King of Denmark, or from both, and thereby to bypass the Staple at Bergen. As early as 1413, as has been seen, an English trading-ship had arrived in Iceland with a licence from King Eirík; and next year no less than five merchantmen had come to the island with letters from Henry V. Fourteen years later a licence was issued for 'John, Bishop of Hólar in Iceland, an Englishman by birth, to buy in England and export either from Lynn, Hull, or Newcastle-on-Tyne, after paying custom, etc., 1,000 quarters of wheat and . . . providing always that letters testimonial certifying the unloading of the same in Iceland be delivered in Chancery before Michaelmas next';[34] and it was in an English vessel that the new Bishop voyaged to Hafnarfjörd on the west coast. In the following decade John, Bishop of Skálaholt, enjoyed the same privilege. In 1436 the Bishop was allowed to send one ship to Iceland and back with merchandise. In 1438 the permission was renewed. In the same year the Bishop of Hólar was allowed to trade with two vessels, and next year with several. Certain merchants of Bristol were also permitted to trade direct to Iceland. In 1439 a licence was granted to Stephen Forster, William Canynges, and Jordan Spryng to take the *Katherine* of 100 tuns or under (master, Peter Gegge) to Iceland and Finmark for fish, the consent of the King of Denmark having also been gained on account of the great debts owing to these merchants from the Icelanders. The year following William Canynges, Jordan Spryng, George Roche, and Richard Alberton were allowed to send to Iceland and Finmark, for fish and other merchandise, the *Mary Redcliff* of 140 tuns or under (master, John Matthew); and in 1442 William Canynges and Stephen Forster were granted leave to dispatch both the *Katherine* and *Mary Redcliff* to Iceland and Finmark for four years. In 1443 no less than nineteen licences were issued to merchants desiring to voyage to Iceland.[35]

Moreover, judging from the numerous cases of arrest, forfeiture of ship and cargo, imprisonment, etc., that we meet with in the national archives during the next decade or so, there must have been many who determined to dispense with the formality of a licence, and to follow the more economical, if riskier, course of sailing direct to Iceland. At any rate, by 1436 it is said that there were so many of them that not a few were unable to get a cargo and had to return home with empty holds.

> And now so fele [many] shippes thys yere there were
> That moche losse for unfraught [without a freight] they bare.
> Yselond myght not make hem to be fraught
> Unto the hawys [havens]; this moche harme they caught.[36]

It is apparent from the King of Denmark's complaint of 1431 that illegal English trading had been going on, on a fairly large scale, for the last twenty years or so—and not only to Iceland, for 'Greenland, the Faeroe Islands, the Shetlands, Orkneys, and other isles' are also mentioned in the complaint.[37] So great was the volume of English traffic, in fact, that there was little left for any ships that chanced to arrive from Norway; and the King's dues fell off in consequence. In the same year that the Danish King's complaint was dispatched, no less than four large merchantmen sailed to Iceland from Hull.[38] (It is significant that their destination is not entered in the English customs accounts as Iceland. That they sailed there, however, is sufficiently clear from the cargoes they carried.) On 24 December 1432 a treaty was signed between Eirík and Henry VI of England.

During the years which followed the English government undoubtedly made some attempt to enforce the ban on the Iceland voyage. In 1436 certain mariners of Kingston upon Hull, Bristol, and Pembroke were required to answer charges arising out of 'divers trespasses in Iceland'. In the same year it is stated that 'a ship and a balynger clepid the *Petus* and the *Julian* of Nucastell with ccxx persones maryners and sowdiours for the safe garde of the See . . . toke a shippe of Bristowe clepid the *Cristofre* the whyche had be in Island and Norway ayenst the fourme of the statute purveid ayenst the contrarie'. In 1439 the *Trinite* of Bristol was forfeited to the Crown for a like offence. In 1448 another large Bristol merchantman, the *Christopher*, which had sailed to Iceland in April with a cargo of cloth, iron, salt, mead, malt, flour, and honey, and brought back 45 lasts of stockfish and 1,800 salt fish, was forfeited to the King. Stockfish from Iceland was seized at Hull and other east-coast ports as well as at Bristol. From the successive proclamations that were made and commissions that were appointed to inquire into infractions of the statute throughout the ensuing years it appears that a large-scale smuggling trade was going on all around the coasts of England and Wales.[39]

The English trade to Iceland reached its height towards the middle of the fifteenth century. The most prosperous years of the traffic were probably those between 1430 and 1460. The larger merchantmen of Bristol, Hull, and Lynn carried general cargoes, the contents of which were sometimes valued at several hundred pounds sterling. For instance, the *Katherine* of Hull (master, John May) left the Humber in June 1436 with a cargo comprising: 20 woollen broadcloths of divers colours, 12 woollen cloths, 38 pieces of linen cloth, 106 barrels of beer, 4 tuns of wine, 10 pipes of wine, 12 pipes of wheat flour. The *Christopher* of Bristol (master, John Page) sailed to Iceland in the spring of 1448 laden with 7 weys of salt, 3 tuns of mead, 18 barrels of osmund, 32 narrow cloths, 11 dozen cloths, 8 dozens of narrow cloths, 6 weys of salt, 7 lasts of flour, 20 barrels of flour, 5 barrels of honey. The *Julian* of Bristol (master, John Stafford) returned from Iceland on 4 September 1437 with '4 lasts gillfish, 7 C titling, 16 C saltfish, 6 C gillfish, 4 lasts titling, 11 M mortes, 5 C hake, C pollack, 4 rolls wadmal, 3 dickers hides, C pollack, $\frac{1}{2}$ C milwell, 1 quarter hake, $\frac{1}{2}$ C hake'. On the same day the *Marie* of Bristol (master, William Crues) returned with '6 lasts gillfish, 13 C titling, 20 C salt fish, 6 waddis wadmal, 9 C gillfish, 6 lasts titling, 17 M mortes, 2 M 4 C hake, 1 M pollack, 9 barrels white herring'.[40]

In a merchant's handbook of about 1500, entitled *The Noumbre of Weyghtes*, there is a brief list of the principal exports and imports of Iceland in this era: 'The cheffe merchaundyse in Iseland ys stokefysih and Wodemole and oyle; and good merchaundyse from hens thedyr the course Ynglysche clothe coloured, mele, malte, bere, wyne, Salettes, gauntlettes, longeswerdes, lynon cloth and botounes of sylver, awmbyr bedes, knyves and poyntes, glasses and Combys etc.'[41]

The Iceland passage was not the only ocean venture undertaken by English mariners at this time. We have already observed how, by 1423, they were voyaging to the Faeroe Islands. In the summer of 1428 two English merchantmen are recorded to have visited Finmark—another great fishing centre. The Danish authorities raised the same objection to the Finmark voyage as they had to the Iceland traffic, and with as little effect. Judging from the complaints made by the King of Denmark, Englishmen had apparently been sailing to Finmark from about the year 1410. It would seem that voyages to that land became fairly frequent. It appears to have been the practice sometimes for ships which had been in Iceland to sail thence to Finmark, instead of returning to their home ports in England. Thus in 1436 it is recorded that the *Cristofre* of Bristol sailed to Iceland and the forbidden parts of Norway.[42]

In the latter half of the century hostilities broke out between England and the Danish–Hanse combination: eventually peace was patched up for a short time, but was ruptured again in 1467, when a band of English merchants slew the governor of Iceland, Björn Thorleifsson, at Ríf on the west coast. His widow appealed to the Danish King; and an ultimatum was dispatched to all English towns save Lynn, the citizens of the latter having apparently been responsible for the outrage at Ríf. The English government thereupon cast into prison all the Hanse merchants in London; and though they were subsequently released a protracted privateering war was carried on between England and the Hanse, punctuated by sporadic and short-lived truces, during which the passage of the Sound was used as a bargaining counter.[43]

Though the English trade to Iceland was beginning to fall off—especially following on, and largely in consequence of, the affair at Ríf—a substantial traffic was still apparently carried on. It is clear that in the 1470s and 1480s English merchantmen, as well as fishing doggers, were still voyaging to Iceland. In 1483 Edward IV granted a licence to Robert Alcock, a merchant of Kingston upon Hull, to dispatch a vessel of 250 tuns, or under, to Iceland. Next year there were merchantmen from Bristol in Iceland.[44] It is evident, too, from the accounts of the quarrels between English and German merchants in Iceland that the English vessels concerned were of substantial tunnage. In 1475 it is recorded that Henry Daniell, governor of Iceland, assisted by some of the Germans, was levying war by sea, *bellum maritimum*, against certain ships of Scarborough. Two years afterwards the *Little Peter* of Hull is mentioned in connection with an attack on certain Hanse merchants in Iceland. For years to come the official records of the Hanse were full of indignant complaints against English merchants—especially those of Hull and Bristol—who for so long had been wont to regard the Iceland trade as their own particular preserve.[45] But in the closing decade of the century English merchant-men of any size seldom made the Iceland voyage. The *Anthony* of Hull, 280 tuns, which was in Iceland in 1491, must have been one of the last of these.[46] Henceforth the English share of the traffic, which by this time had much declined, was carried on by doggers.[47]

19

The Iceland fishery

A MAJOR FACTOR in English maritime enterprise at this time was the fisheries. The lesser fisheries of the Channel were worked mainly by the Cinque Ports, who also sent a large fleet each year to the North Sea for the autumn herring fishery and assisted at the ancient fish-fair of Yarmouth, the overlordship of which belonged to the ports. Certain fishing-ports had a particular reputation for a certain fish. We hear, for instance, of the *Haraung de Gernemue* (Yarmouth), *Playz de Wychelsee*, and *Merlyng de la Rye*.[1]

The opening years of the fifteenth century saw the eclipse of the rich herring fishery at Skania—whose salted and smoked herrings were famed throughout Europe, and which had contributed in such large measure to the strength and prosperity of the Hanse—as the herring shoals moved out of the Baltic into the North Sea. From this time on the herring fishery off the east coast of Britain became the most important in Europe. It was worked about equally by the English and the Dutch. Another highly important fishery in home waters was that off the south and west coasts of Ireland, which was much frequented by vessels from our western ports. In the late medieval era our fishermen began to go much further afield.

From the first decade of the fifteenth century the Iceland fishery was systematically exploited by the English. Year by year a large fishing fleet—drawn, for the most part, from the east-coast ports—made the passage to Iceland. Later in the century it is stated that they would sail from England early in the year—in April, March, or even February—fish for about three months off Iceland, and return home again in July, August, or September.[2]

Some of them fished off the north coast in Skagafjörd; but more often they worked the great flat which fringes the whole west coast of the island. Their principal centre, perhaps, at any rate during the early years of the traffic, was the Vestmannaeyjar, lying off the south-west coast. Here was said to be the best fishing in all Iceland, 'in illa insula est melior piscatura tocius ijslandie'.[3] A glance at the chart will suffice to show why these islands were—and are—so greatly important from the navigational viewpoint. Situated off the dangerous southern coast of Iceland—its exposed and surf-lashed strands affording no safe shelter for many miles in either direction: frequently obscured by mists and dust: fringed, at numerous points, with outlying reefs, rocks, and shoals—the Vestmannaeyjar not only offered a tolerably secure anchorage, but also lay in close proximity to the fishing grounds.[4]

Already by 1420 some of the English were wintering in the islands.[5] Each year the disorders in Iceland were increasing. Not a few of our fishermen were a truculent, lawless crew and had given trouble almost from the start. These depredations in the next decade were set out in a comprehensive indictment by the governor, Hannes Pálsson, of 'the men of Hull and other English towns'.[6] How for several years they had been coming to Iceland, contrary to the law and ancient custom of the realm, and how the King of Norway was thereby cheated of his dues; how they had slain a number of the royal officials in Iceland and wounded several others, besides taking many prisoners, and despoiling them of their goods; how they had robbed and looted and burnt churches; and how vessels from Norway could not safely sail to Iceland, nor remain there, for fear of 'the men of Hull and other English towns'. The tactics of certain of these east-coast mariners are strangely reminiscent of the proceedings of William Hawkins and his contemporaries some hundred and fifty years later in Spanish America. Truth to tell, 'the men of Hull and other English towns' were ruthless and determined rapscallions who were out to get the stockfish for which they had voyaged so far by fair means or foul. What Chaucer says of his Dartmouth Shipman is no less true of these east-coast fishermen, 'Of nyce conscience took he no keep'.

In 1422 English marauders were embroiled with the royal officials at Bessastadir. A skirmish broke out in which a man was fatally wounded and a great deal of damage done. Next year the men of Hull were once more to the fore (Lynn is also mentioned in the report). Mention is made of raiding and plundering and burning of churches, especially around the Ólafsfjörd. Yorkshire names again occur in connection with English depredations in 1424.[7]

The report further declared that every year the English steered for the Vestmannaeyjar, where they were beginning to erect lodges, till the soil, and use these isles, in fact, as if they were their own property. 'They have neither sought for nor ever obtained permission from the Icelandic officials, but occupy the place by force, and do not permit the fish belonging to the king or any other persons to be shipped from the island before they have loaded their own vessels according to their pleasure.' So long as the English remain in Iceland, the report continued, the people cannot go to sea or carry on their fishing, for the English smash their small skiffs (scaphas), and 'beat up, wound, and otherwise maltreat' the native fishermen.[8]

Two important facts emerge from this report. First, that the quarrel was mainly between the English and Danish officials in Iceland who were vainly endeavouring to uphold the King's rights and to exact the King's dues, rather than between the English and native Icelanders. Second, that the lawless elements among the English sailing to Iceland were—judging at least from the names cited in Hannes Pálsson's report—chiefly from the east coast. At the same time due allowance must be made for the fact that the author of this indictment was in no way an unbiased witness. It was obviously his intention to expel the English from Iceland; but in this he was unsuccessful.[9]

In 1425 things reached a climax. In that year the governor and another official, Balthasar van Dammin, had rather rashly crossed over to the Vestmannaeyjar to arrest the English who were established in the islands and to confiscate their ships. The latter put up a determined resistance. The consequence was that the English not only beat off the attack, but destroyed the boats in which Pálsson and Balthasar had come over from the mainland, thus forestalling their escape. The two officials were then overpowered, captured, and carried off to England: a departure which, as the Icelandic annalist drily observed, 'few regretted'.[10]

Once again King Eirík straitly forbade all foreigners to sail to Hálogaland,

Finmark, Iceland, and 'our other dependencies'. In September 1426 the veto was proclaimed in Bergen in terms which went straight to the root of the matter— 'super caussis et articulis transgressionorum anglicorum de Hull [Hull as usual heads the list] Eboraco et lenn et alijis locis anglie jslandia'.[11]

Little is known about Hanse activities in the Iceland fishery; but there is no sign or suggestion that the English lost ground to German fishermen as they presently did to German merchants. On the contrary, the fishery appears to have remained almost exclusively in English hands. A far more serious handicap, indeed, in the case of many of the east-coast fishing-ports than the rivalry of the Hanse and the raids of foreign and home-bred pirates seems to have been the condition of their havens and channels, which were in continual danger of destruction. Several of the most flourishing havens of East Anglia were now threatened with a fate similar to that of the Cinque Ports. In the course of the fifteenth century repeated, and in the long run unavailing, attempts were made to keep open the harbour of Dunwich, which was continually filling with shingle. Southwold and Walberswick were destined in this way to lose both their haven and their livelihood. A local jingle put the matter in a nutshell:

> Dunwich, Sowl, and Walderswick
> All go in at a lowsy crick.

There is good reason to believe that in some years the fishing fleet amounted to as many as a hundred vessels. This estimate has to be arrived at by a process of deduction from various sources, as exact figures are seldom available. During the last two decades of the fifteenth century the English traffic to Iceland became more or less regularized. To this period belongs an interesting example of trade protection. When armed Hansemen began to ply to Iceland, the English government took steps to protect the fishing fleets by the provision of convoys. The first mention of these occurs in the reign of Richard III.

Ricardus, &. To all maner awners, maisters, and mariners of the naveye of our counties of Norffolk and Suffolk, aswele fisshers as other, intending to departe into the parties of Island, and to every one of them, greting:—Forasmoche as we understande that certain of you entende hastely to departe towards Island, not purveied of waughters for your suertie in that behalve; we, for certain grete causes and consideracions us moving, woll and straitly charge you, alle and every of you, that ye noon of you severally depart out of any of our havens of this our realme, towards the said parties of Island, without our licence furst had soo to do: and thereupon, that ye gadre and assemble your selff in such one of our havens or portes in our said counties of Norfolk and Suffolk as ye shall thinke most convenient, wele harnyssed and apparelled for your owne suretie, and soo forto departe alle togider toward Humbre, to attende there upon our shippes of Hull as your waughters, for the suertie of you all; and that ye keep you togeder, aswele going into the said parties as in your retorne unto this our realme, without any wilfull breche to the contrarie, upon payn of forfaiture of your shippes and goodes in the same.[12]

The traffic to Iceland was now an annual venture, regulated by royal ordinance and stature. At the same time the trade was undergoing a significant transformation. The comparatively small craft known as doggers, which had for so long and in such considerable numbers been plying to Iceland, were about to monopolize the English share of the traffic entirely. Substantial merchantmen from Hull or Bristol are seldom or never heard of in these later years. Their place is filled by the triumphant Hansemen. Nevertheless, though the large German trading-fleets absorbed most of the commerce of Iceland, the English could continue to work the fisheries, and so fill their holds. The doggers could also take out with them a certain amount of merchandise; but the export of grain, as may be seen from the following letter addressed by Henry VII to the Lord Admiral in 1491, was forbidden to the doggers.

Ryght trusty and wellbelovyd cousyn, we grete you well, etc. In that ye desyer all the dogers of those partes schuld have our licens to departe in the viage towardes Island, as they have ben accustommyd to do yerly in time passyd, and that ye woll undertak they shall have with them no more quantities of graynes then woll only suffice for ther vitallyng and expensis; we lete yow withe that our fully interly belovyd cousyn and Kyng of Demarke, hath showyd and compleynyd un to us by dyverse his letters, that when our subjectes ther ageynse ryght and conciens. Wherfore, the seyd doggeres fyndyng sufficient surte be forne yow, such as ye will answer unto us, that they shall not have with them no graynes mo then shall only suffice for ther vitallyng, nor odyr thyng woth them that ys for bedyn, and also that they shall not in goyng, comyng, nor in ther beyng at the seyd Island, take noo thyng but that they treuly pay or agre for, and frendly entret our seyd cousyns subjectes withowth eny robbyng or ex'startyng them in there bodyes or goodys; we be content the seyd doggers make their viages thedyr at ther libertes, eny our wrytyng or comandment med to the contrary not withstandyng; and allys we woll that our restraynte of ther thedyr goyng stond styll in his strenthe and vertu.[13]

The Iceland fishery prospered exceedingly throughout the fifteenth and early sixteenth centuries, reaching its peak period, perhaps, in the course of the decade 1520 to 1530.[14] We are indebted to Gardner's *Historical Account of Dunwich*, a work which is apparently based on various local sources now unfortunately lost, for a mass of information concerning the East Anglian traffic with Iceland during this time. Gardner relates that in 1451 Walberswick alone possessed no fewer than 'thirteen Barks trading to *Iceland*, *Farra* (Faeroes), and the North Seas: And on their own Coast, twenty-two Fishing Boats for full and shotten Herrings, Sperlings, or Sprats, &.' In the same work there is printed the will of William Godell, first bailiff of Southwold (dated 1509), which gives some idea of the proportions which the Iceland fishery had by this time assumed.[15]

According to the leading authority on the English trade with Iceland, Björn Thorsteinsson, 'It is clear from the sources that English vessels sailed to Iceland to trade and to fish in greater numbers in the first decade of the sixteenth century than at any other time down to the nineteenth century.'[16] By the year 1528 there were nearly 150 English craft engaged in the Iceland fishery, the great majority of which belonged to the east-coast ports, and especially to those of East Anglia. Thus in an early sixteenth-century document of the High Court of Admiralty, a vessel of Yarmouth is mentioned as 'Rygged with all hyr Salte vitell and apparell to hyr perteyneth for the viage to Iislond and preste to goo a Dogger fare'. The Southwold Bay group of fishing-ports, which came to the fore about the middle of the fifteenth century, were now at the height of their prosperity. In a lengthy but badly mutilated report to Henry VIII in the year 1533 the King is informed that 'every year there is maintained and rigged for the said . . . Norfolk and Suffolke and other parts of this your realm of England 100 sail . . . above that do go of merchandizes and "dogger-fare" for ling, cods and stockfish'.[17]

In 1490 Hans I granted Englishmen the right to trade and fish in Iceland so long as they had the King's licence. But the *Althingi* disallowed the clauses granting fishing rights to the English, and in the ensuing years only permitted them to fish in Icelandic waters if they imported goods necessary to the island and took fish in exchange. These provisions had little practical effect on the course of English traffic with Iceland. In 1500 the *Althingi* protested that the English, lying in the offing with their larger craft and long lines streaming down the tide, were spoiling the inshore fishing for the Icelandic fishermen in their smaller vessels. But in view of the fact that this complaint was reiterated more than thirty years later it would appear that things went on more or less as before.[18]

As has already been emphasized, towards the end of the fifteenth century the large English merchantmen which had formerly traded to Iceland were gradually supplanted by relatively small doggers, which, judging from the nature of the cargoes they carried, clearly engaged in commerce as well as in fishing. Firkins of butter, bales of woollen cloth and linen, copper kettles, caps, knives, 'gerdelles et pynnes', 'poyntes nedilles et laces', were among the goods they transported to Iceland, and these, in the aggregate, in considerable quantities. 'The people be good fishers,' observed Andrew Borde; 'much of theyr fishe they do barter with English men for mele, lases, and shoes & other pelfery.'[19] It is stated that some of these doggers would carry between 40 and 50 tuns of Gascony wine, which gives us some idea of the approximate tunnage.[20] Some, however, were much larger than this, while others were a good deal smaller.

In a number of documents belonging to the early half of the sixteenth century a revealing light is thrown on the fitting-out and victualling of these craft bound for the Iceland fishery. In the first of them, which is taken from Henry Tooley's Day-Book, inscribed *By the Grace of God to Icelandward 1526*, we learn that a ship's stores included such items as: '32 hundred hooks at 6d. the hundred; 1 firkin with gunpowder $32\frac{1}{2}$ lbs. at 6d. a lb.; 5 bushels of white peas at 8d.; $\frac{1}{2}$ hundred leyngs, 30s.; 1 toppnett figs [hempen net of figs], 2d.; 2 barrels of tar, 1 barrel of pitch, 20s.; 1 barrel beef, 10s.; 8d. for gargyng hooks . . .' In 1534 the provisions, etc., taken on board included: '18 wey of gray salt; 38 barrels of beer; 50 couple of salt; 36 pipes of beer; 18 flitches of bacon; 2 old cables; 4 new cables; 6 anchors; 2 doz. bull anchors; 20 pyens and 4 danes; 1 barrel of beef; 1 firkin of butter; 84 lbs. of cheese 16s. 0d.; 1 barrel of weythe hern [white herring]; 1 cade of red herne; 1 boat rope.'[21]

Two years later there appears the 'Bargain of Thomas a Bury and his company, 1536; by the Grace of God to go to Iceland in *The Mary Walsingham* and he shall have 18 wey of salt and he must have to every wey of salt, a quarter of wheat, two pipes of beer, two flitches of bacon . . . one great pice of ordnance and two serpentines four hackle-bushes [harquebuses] one great meat kettle and one mynggyng kettle, one pitch kettle, and one crow of iron and a great hamber, ten bows and 2 sheaves of arrows and a firkin of gunpowder.'[22]

Lastly we have a document bearing the endorsement *A Raconyng of a Vyage into Eyeslond 1545*: 'The Cargis of a good schyppe callyd the *Jayms* of Sir Thomas Darssys knyght for a viage be the grace of god to be mayd yn ysland begonne the fyrst day of Dessember anno 1545 mayster vnder god for thes present viage long Sander of Sunwhyche & marchant Jeffrey Smythe'. On board the good ship *James* of Dunwich the stores included: '15 wey & a hauf of Salt at 36s. 8d. the wey; guttyng knyefes; 3 splyttyng knyeffes; strynges to the number of 40 dossyn at

6s. 8d. the dossyn wythe chargis; tares 7 dossyn at 9s. 8d. the dossyn; haulf a dossyn of skoypes; 2 Lanterns; 2 Boet Compassys; hoykes at 6s. the 1000, 15,000; all kynd of nayls for store', etc.[23]

The importance of the Iceland traffic in the national economy of the period under review is revealed by several significant references in the English state papers. For instance, in 1523 Surrey told Wolsey of a report that the Scots were intending to send out six or seven ships to the Orkneys, to intercept the Iceland fleet on its way home. 'If they succeed, the coasts of Norfolk and Suffolk will be undone, and all England destitute of fish next year.' Soon after Wolsey in a letter to the King mentioned that the Iceland and Zealand fleet, and among them the *Mary James*, the richest, had returned safely. Steps were taken to escort the fishing-fleet safely clear of the Northern Isles. At the end of the season similar precautions were taken on the return passage. In Icelandic waters, however, no adequate protection was available.[24]

20

The English rivalry with the Hanse

IT WOULD APPEAR that the Hanse first launched out on the ocean trade-routes a decade or so later than the English. According to the official records of the Hanse, the constituent members of the League in 1416 undertook to forbid their merchants to sail to the Orkneys, Shetlands, and Faeroe Islands. It would seem that the Hanse began to venture to Iceland in about 1420, probably from Bergen. In an edict by King Eirík in 1425 there is a reference to Germans and other foreigners who sailed to Hálogaland, Finmark, and Iceland. Six years later the Althing decreed that neither English nor Germans were to winter in the island.[1]

Presently Dutch as well as German merchants and mariners joined in the trade. In the same year that the Althing issued their decree the *Holy Ghost* of Schiedam left the Humber for Iceland with a cargo of English merchandise for Iceland. In 1439 the *Marieknyght* of Amsterdam sailed from Iceland with a cargo of stockfish and other goods for Dalkey in Ireland. There were probably other Dutch sailings of which the record has been lost; since in 1443 Christopher of Denmark forbade the mariners of Brill and Schiedam to voyage to 'Iceland and other dependencies', *Ísland og önnur skattlond*.[2]

In the official records of the Hanse the Iceland trade is first mentioned in the year 1434–5.[3] At this stage the new venture met with strong opposition within the League itself; and it is not too much to say that for over half a century English merchants trading to Iceland encountered no serious competition.[4] They had profited by the war which broke out between Denmark and the Hanse sea-ports in the early part of the century. But peace was restored in 1435, and the Danish kings fell increasingly under the influence of the Wendish group which dominated the councils of the League and monopolized the trade of Bergen. Towards the close of the 1460s the Hanse were at last able to present a united front to their rivals, whose position was, moreover, undermined by the vacillating policy of the English Council. For political reasons it was impossible during the troublous times of the baronial wars, when the English throne was the prize of contending factions, to offend the all-powerful League, by means of whose assistance Henry IV had landed to dispossess Richard II in 1399, and Edward IV had made his bid for the throne in 1471. Furthermore, on the breakdown of the Lancastrian government the English merchants could expect no support against their enemies, and, in any case, English acts of piracy on the high seas had repeatedly played into the hands of the Hanse.

At the same time the English trade with Iceland was entering on a period of retrogression. The turning-point in these developments may be said to have been the murder in 1467 of Björn Thorleifsson, the governor of Iceland, referred to in an earlier chapter. By way of reprisal the Danish King seized a number of English merchantmen in the Sound. The connection between Denmark and the Hanse was now so close that a quarrel with the former involved, almost automatically, hostilities with the latter. In the ensuing war between Denmark and England, the North German Hanse joined in on the Danish side. The Danish control of the Sound was the dominant factor. 'It is true that the Danish cannon at Copenhagen did not cover the Iceland seas,' observes Thorsteinsson, 'but it was in effect due to Danish supremacy in the Baltic Straits that Iceland remained a possession of the Danish Crown.'[5] After a maritime war in which English trade suffered severely, peace was restored in 1473 on terms very advantageous to the Hanse. In the mid-seventies the German share of England's foreign trade was larger than it had ever been before; and the English traffic to the Baltic was virtually at an end.[6]

The same period which saw the triumph of the Hanse in the North Sea and the Baltic witnessed also a steady expansion of the German trade to Iceland. In the course of the 1470s German competition began in earnest. The English merchants at first vigorously resisted the encroachment of the Hanse; between 1474 and 1532 recurrent clashes, of increasing intensity, broke out between the old-established traders and the new.[7]

The official records of Hamburg bear witness to the rapid expansion of the German traffic with Iceland in the final quarter of the century. In 1476 there appears an entry, 'de naulo sive affretatione expeditarum versus Islandiam', in the course of which mention is made of two ships, the *Hispanigerd* and *De Grote Marie*, which had made the voyage to Iceland the year before. This would seem to have been the start of the regular trading voyages from Hamburg to Iceland. Both ships belonged partly to the city of Hamburg. The trade to Iceland was a civic venture, financed from public funds. From now on the Hamburgers sailed annually to Iceland, and the city kept a profit and loss account of these ventures. In the year 1490 there were no less than five ships from Hamburg in Icelandic waters. Despite the fact that the Iceland voyage was still formally prohibited by the Hanse council in the interests of the Bergen *kontor*, the astute Hamburgers found ways and means of circumventing the veto. In 1484 the latter had freighted a vessel that was being laded at Wismar for the Iceland voyage; and subsequently the reply of the emissaries from Hamburg was that, to the best of their knowledge, 'no ships had been sent from Hamburg on the Elbe'.[7]

Though, as is evident from the protests of the Bergen *kontor* to the Hanse council at Lübeck, the traffic was mainly in the hands of the Hamburgers, it is certain that, in the second half of the century, several other of the Hanse sea-ports were engaged in this commerce, namely, Bremen, Danzig, and Rostock. Later Stralsund, Luneberg, and Wismar also joined in the trade, which in the 1480s and 1490s appears to have rapidly expanded, and to have embraced the Shetlands and Faeroes as well as Iceland. Nor were the Germans the only rivals the English merchants had to face in the later decades of the fifteenth century. The Dutch were again sailing to Iceland. In 1471 two ships from Amsterdam are recorded to have arrived in the Hafnarfjörd; and in 1490 the Iceland venture was formally opened to Dutch shipping, more especially that of Amsterdam. But Holland's share of the traffic never amounted to anything much, and the Dutch dropped out of the trade altogether in the early half of the sixteenth century.[8]

It would be a wearisome and, indeed, unprofitable task to probe into the rights and wrongs of a controversy that ran its course, with mutual injuries and recriminations, for more than half a century. Almost from the start, it was less a feud between the races than a fight between factions. The English fought with the Germans, as well as with the Danes, both by land and by sea: the Germans contended among themselves: there were complicated and endlessly changing combinations and coalitions. In their anxiety to monopolize the Iceland trade, the Hanse were ready and willing to assist the Danish officials against the English in return for increased commercial concessions. In the end the German Hanse largely supplanted their rivals in the Iceland trade, though not in the fisheries. When King Hans in 1490 legalized the direct trade to Iceland, it was with the hope of holding the Germans in check by encouraging the English and Hollanders (especially the men of Amsterdam) to sail to Iceland. But little came of this measure. The decree which forbade foreigners to winter in the island, which had been so many times re-enacted, had become practically a dead letter. In the sixteenth century the Germans in fact overawed the Althing; and in 1531 the Archbishop of Nidarós complained that the Hamburgers held Iceland and the Faeroe Islands 'as fiefs'.[9]

Exactly at what date the broad-beamed, bowl-shaped, single-masted cog [10] of the North German and Netherlands ports gave place to the two-, three-, and four-masted sailing-vessel cannot be determined with certainty. But there is no reason to suppose that the change was made on the other side of the North Sea any earlier than it was over here. There is no evidence of an improved sail-plan in Germany or the Low Countries during the first thirty years or so of the fifteenth century. Down to the year 1400 the city seals all show a single-master. Nor are the exact points of difference between the Hanse cog and the Hanse hulk by any means clear.[11]

As in England, the advent of the three-master in continental Europe seems to have occurred about the middle of the century. In 1462 a 'great caravel', called the *Peter* of La Rochelle, is recorded to have visited Danzig. The *Peter* was fitted with foremast, mainmast, and mizenmast. About the same time, on a seal belonging to Louis de Bourbon, Admiral of France, there is a representation of a square-rigged three-master. Another striking example is the Flemish *kraerk*, or carrack, portrayed by the painter 'WA'. This is a large, strongly rigged three-master fitted with a lateen mizen.[12]

Though earlier evidence is lacking, it is apparent that similar developments had been taking place in the German Hanse ports. To the late fifteenth century must be ascribed a series of crude sketches of what appear to be three- and four-masted Hanse sailing-vessels, carrying lateen mizens, scratched on clay bricks in the Carmelite monastery at Elsinore. Among these sketches is also one of a craft bearing a striking resemblance to the *kraerk* of 'WA', which was apparently a Flemish carrack. Furthermore, at St. Mary's church at Lübeck, where the Company of Bergen Traders had their chapel, is to be seen a painted panel (dating from the year 1489) depicting a number of large sailing ships flying the red and white colours of Lübeck, carrying three and four masts. The largest of these craft is fitted with the second, or bonaventure, mizen that was coming into vogue towards the close of the century.[13] It is possible from these pictures to form some conception of the general appearance of the larger Hanse merchantmen sailing to Bergen and Iceland.

The slow-sailing, heavy Hanse cogs would reach Bergen, by way of the Great Belt, from Lübeck in some three or four weeks. The voyage from Hamburg to Iceland took rather longer. The vessels of the German Hanse usually sailed in convoy, escorted by an armed cog, *vredecogghe*.[14] In this connection it is to be noted that the painting in St. Mary's church, Lübeck, of the great, 'high-charged' Hanse three- and four-masters, powerfully manned and bristling with guns, serves to explain how it was that the German Hanse were able at this time to overcome their English rivals in the struggle for the Iceland trade.

Iceland-bound merchantmen of Hamburg sailed as a rule in March or April (the earliest recorded date of departure is 11 March); the return passage was made in July or August. The Hamburgers evidently sailed for Iceland rather later than did the English; and the reason for this, it has been plausibly suggested, was that the Hanse merchants were interested only in trade, while the English were also concerned with the Iceland fishery. No commerce was permitted in Iceland before 1 May; and it follows that there was no advantage to be gained by arriving in the land at an earlier date. As to the duration of the direct voyage between the North German ports and Iceland, there is not enough evidence to arrive at anything like a precise estimate; but it was conjectured by Baasch that the passage probably took about a month.[15]

From about the year 1400 fairly frequent references to the mariner's compass and lodestone are to be found in the records of the Hanse. In 1394 mention is made of certain 'compass-makers', *konpassemakers*, in the records of the confederation. Six years later a dozen compasses are stated to have been shipped to Yarmouth from the Low Countries. In 1460 a Danish vessel, lying at Marstrand in the Baltic, was equipped with 'compasses and sailing-stones'. In the Hamburg archives in 1461 there is a reference to a payment made by Gerard of Essen for two compasses, and, ten years later, for 'a compass and night-glass'. In 1475 a vessel of Amsterdam was provided with 'compasses and glasses', and in 1478 a craft from Oslo was equipped with 'fishing gear, lead and lead-line, compasses, and all her gear'.[16] The often-quoted observation made by Fra Mauro to the effect that neither chart nor compass was used in the navigation of the North Sea and the Baltic ('per questo mar non se navega cum carta ni bossola, ma cum scandaio')[17] is not to be taken too seriously. From the fifteenth-century German rutter, *Das Seebuch*, numerous compass courses may be cited for both the seas in question. Like the English, the mariners of the German Hanse steered by the compass in conjunction with the lead and line. How supremely important was this latter item of the ship's gear may be seen from the account given in the Hanse archives of a Danziger, bound for Lisbon, placed under arrest at Plymouth in the autumn of 1449. To make certain that she did not slip out, the local authorities had the ship's lead and line (*lyne und loth*) taken ashore.[18]

Das Seebuch is silent as to the course followed by Hanse merchantmen trading with Iceland. However, in 1469 a vessel of Bremen is recorded to have sailed out of the Weser with a general cargo, bound for Iceland. She sailed apparently by way of the Shetland Islands (where she ran aground): a route which is mentioned in the Dutch edition of *Dit is die Caerte vander See cost ende west te seylen*, published in Amsterdam in 1566, but based on an original of earlier date. It is likely that other Iceland sailings of the German Hanse followed the course set out in this, and the English version of the work, entitled *Safegarde of Saylers, or great Rutter*.[19]

Finally, it is apparent that even with the steady advance of shipbuilding and navigation the passage to Iceland could still present difficulties. It is on record that one 'scipper Horneman' was unable to make Iceland and was eighteen weeks at sea—'De hadde 18 weken in der see gewest und hadde Island nicht finden kont'.[20]

During the last part of the fifteenth century the Germans developed the export of a number of Icelandic commodities which the English, in the main, had neglected, including *váðmal*, hides, train-oil, sulphur, falcons, and eider-down. The Hamburgers exported to Iceland large quantities of foodstuffs, including corn and beer. So much corn, in fact, was shipped from Hamburg that it created a dearth;[21] and in 1483 the trade was prohibited by the city council. In the same year a vessel of Lübeck is recorded to have carried flour, butter, and ale to Iceland. The ability of the Icelanders to pay for all these imports from Europe may be accounted for by the relatively prosperous state of the country during the fifteenth century.

It is difficult if not impossible to treat of the Hanse trade except in the most general terms. No really comprehensive statistics are available regarding the volume of this trade or of the number of vessels engaged in it. There are, however, certain passages in contemporary chronicles and archives which enable one to form some conception of the ample resources of Hanse shipping. Thus it is recorded that, in a September gale in 1491, no less than ninety German vessels were wrecked on the shores of Iceland; some 3,000 dead bodies, it is said were later washed ashore.[22] Even allowing for some measure of exaggeration it is evident that the German traffic to Iceland was by this time on a very large scale indeed.

One point, however, needs to be stressed: the Germans were hardly of the same nautical calibre as the English and Hollanders of the same day, or as the Norwegians of an earlier period. The large fleets of small fishing doggers which sailed every year for the 'costes colde' from our east-coast ports give testimony of abundant reserves of prime seamen. The standard of seamanship in the English fishing-fleets must have been high indeed for them to accomplish the long ocean voyage so regularly and to face the manifold hazards of the deep-sea fishery. Nor did the German Hanse show anything like the same spirit of enterprise as did the merchants of Bristol, who in the eighties and nineties dispatched successive expeditions into the Atlantic in search of distant islands. The strength of the German Hanse lay rather in their ample reserves of shipping, their accumulated capital and experience, the superiority of their business methods and organization, and their hold upon so many lucrative trade-routes. For nearly two centuries they had dominated the commerce of the North Sea and Baltic. But with the waning of the Middle Ages the political and economic strength of the Hanse was gravely undermined. It was confronted by increasing internal and external difficulties, the cumulative effects of which were decisive. By far the most serious internal problem was the virtual secession of the Netherlands communities. These cities, which had formerly been members of the League, now became formidable rivals. Despite its superior tunnage and old-established position in the North Sea and the Baltic, the League found itself losing more and more of the carrying trade of northern Europe to the upstart Netherlanders. The imposing façade of power and permanence which the League had exhibited for so long rested, in truth, on very unsound foundations. In the maritime history of England the close of the fifteenth century may properly be regarded as the end of the beginning. But in that of the Hanse it was rather the beginning of the end.

21

The last ventures to Greenland

AFTER THE EARLY years of the fifteenth century the history of Greenland becomes more and more shadowy and vague. We learn something from the surviving documents of the period and a good deal more from the reports of Danish archaeologists. It is, in fact, mainly through the admirable work of the latter that it is now possible to carry on the story in outline down to the end of the century.

There is certainly no truth in the allegation that sometimes used to be made that, in the last part of the fourteenth century, communication with Greenland was 'practically abandoned'. For the best part of a hundred years after there appear to have been at any rate occasional sailings to that distant country. The evidence of archaeology serves to suggest that, even towards the close of the century, the Greenlanders were still in close contact with other Europeans. Over and above the archaeological evidence, there are a number of documents relating to conditions in the latter part of the fourteenth, and in the early years of the fifteenth, century, which certainly do not suggest that the colony was then on the verge of extinction.

The first of these documents records certain proceedings brought against a party of Icelanders who, it was stated, had lately made the voyage to Greenland forbidden by law ('att ingen mader skulu sigla i kaupferder . . . til skatlanda waro') and had trafficked with the inhabitants without having gained the King's permission ('att þeir hafdo köft ok selt a Gronlande vttan orlof konungsdoomssens'). The Icelanders' defence was that they had been driven to the coasts of Greenland by stress of weather ('at þeer varo i havdo i storum vanda ok vada ok liifshæska); and they denied that they had broken the law by trafficking with the Greenlanders, who had a law obliging any 'Eastman' who purchased food in their country to take other Greenland products as well. It would further appear from this document that towards the end of the fourteenth century the assembly, or *Thing*, was still held and that a steward (*umbodzman*) was appointed to watch over the King's interests in Greenland.[1]

In the year 1406 another party of Icelanders, who were likewise bound for Iceland from Norway, were driven off their course to the Greenland coast. They were obliged to spend four winters there before they were able at last to return to Norway, and thence to their own land.[2] The intelligence which they brought back with them to Europe is the last certain knowledge that has come down to us concerning conditions in the Eystribygd: and once again there is no suggestion that the colony was in imminent danger of extinction.[3] The sanctions of the law were at any rate still in force there, judging from the death sentence pronounced in 1407 on a Greenlander named Kolgrím who had been found guilty of seducing another man's widow by means of 'the black art' (*svarta kvonstr*) and who was condemned to be burnt alive.

There has besides been preserved a most interesting and unique document: namely, a marriage certificate issued 'at Gardar in Greenland' and dated 19 April 1409. It is the one and only document issued in Greenland which has come down to us. This contains nothing at all that could be construed as proof that the Greenlanders were falling away from Christianity, or that matters in the colony were approaching a crisis; but rather the contrary.[4]

In this document mention is made of two priests, one of them, Sira Eindrid Andresson, who was the bishop's deputy, and the other, Sira Páll Halvardsson, as having officiated at the marriage of Thorstein Ólafsson and Sigrid Bjarnadottir, 'on the Friday after St. Magnus's day', in the church at Hvalsey. It is stated that the banns had been called on three consecutive Sundays, that there had been many people present in church, both Greenlanders and Icelanders, and that everything had been done in accordance with God's law and that of holy church.[5] So that this much, at any rate, is clearly established: in the early years of the fifteenth century, though the land had long been without a bishop, there were still priests living in Greenland.

In 1410 the Icelanders were at last able to get away. 'That year there sailed from Greenland Thorstein Helmingsson, Thorgrím Solvason, Snorri Torfason, and others in a ship bound for Norway.'[6] It is to be presumed that since they made the direct voyage to Norway, instead of sailing to Iceland, their own vessel must have been lost; and that they had had to wait all these years, before a ship arrived from Norway.

The year 1410 witnessed the last recorded voyage from Greenland to Norway. After this there is silence. Official communications appear to have ceased entirely; but two letters from the Holy See, dated 1448 and 1492 respectively, throw some light on the fate of the unfortunate Greenlanders, now believed to be in a very bad way. The first of these (Nicholas V to Bishop Marcellus) told of savage attacks launched by the Eskimo on the Greenlanders, and of Christians, both men and women, led off into servitude. Despite certain details which are manifestly apocryphal, this letter, taken in conjunction with the other evidence, is perhaps worthy of more careful study than it has hitherto received; and will be examined in detail at a later stage. As regards the other letter (Alexander VI to the Archbishop of Nidarós), we stand on firmer ground. The Pope declared that for eighty years or thereabouts the country had been without either bishop or priest, for which reason many had abandoned the faith; that all that the people had left to remind them of the Christian religion was a *corporale* on which, a hundred years before, the Body of Christ was consecrated by the last priest left alive in the land. Severe ice-conditions were referred to as the cause of the extreme rarity of the voyages to Greenland; and the belief was expressed that, for eighty years, no vessel had visited there. Nor, it was added, should a vessel ever make the attempt, could it possibly reach that land save in the month of August, when the ice melted.[7]

It would seem that the increasing difficulty and danger of the navigation to Greenland was an important factor in its discontinuance. The fjords were now blocked with ice during the greater part of the summer. Sometimes the ice did not disperse from one year's end to another, and navigation was impossible. Thus at the very time when the worsening conditions of life in Greenland must have been rendering the lot of the inhabitants almost intolerable, contact with Europe was in danger of ceasing altogether. The Church also had lost touch with the diocese 'at the world's end'. Though successive bishops of Gardar continued to be appointed down to the time of the Reformation none of them ever came out to their diocese.

That intercourse with Europe was not entirely broken off is suggested by certain passages in this letter which appear to reveal some knowledge of the state of affairs in Greenland subsequent to the last recorded voyage in 1410. The reference to the *corporale* in particular is so circumstantial that it may well be due to authentic intelligence. This impression is corroborated and strengthened by a study of the archaeological evidence. Among the finds of later date in the churchyard at Herjólfsnes are certain fragments of clothing which suggest that the Greenlanders were in occasional contact with Europe long after the severance of official communications with Norway. Some of these finds can be dated with a fair degree of precision. They show that almost down to the end of the fifteenth century the Greenlanders were attired in the fashions of contemporary Europeans.

Since the Greenland *knörr* no longer sailed from Bergen as of old one can only conjecture that occasional contact was maintained with Europe through certain unknown foreign vessels calling at Herjólfsnes and elsewhere in Greenland. These foreign vessels may have been either English or German; or possibly both put in there: but the balance of probability is that, if this were so, it was the English who, as in the sailings to Iceland, were first in the field.[8]

The question has sometimes been asked, had English mariners any knowledge of the old Norse colony of Greenland? Though there is no direct evidence on the point, the conclusion is almost inescapable that they surely must have had such knowledge. There are two places at least to which they resorted where there were obvious opportunities for acquiring it. The first was in Bergen, for centuries the headquarters of the Greenland trade; and the second was in Iceland—especially the havens on the west coast. Ships from Bristol would sometimes lie for two or three months in summer in the Breidafjörd, whither a Greenland craft had come in from Markland in the previous century, and later Björn Einarsson Jórsalafara and his party arrived after their four years' sojourn in Greenland. Here the English would get tidings of Greenland along with the other news. In short, among the merchants and mariners of Bergen, and even more among the Icelanders, the general situation in Greenland about the year 1400, must have been a matter of common knowledge. From these likely sources the English may well have learned a good deal about conditions in Greenland and the opportunities for trade there. At the same time they may have become acquainted with the sailing directions for the colony across the ocean.[9]

In the Viking age, as is well known, sundry important discoveries had resulted simply from vessels having been forced off their course and driven before a gale to within sight of an unknown shore. There is every reason to suspect that this pattern of drifting, fortuitous discovery, and planned voyage may have repeated itself in the late medieval era. From the first decade of the fifteenth century large numbers of English fishing doggers had been sailing regularly to Iceland. They had been accustomed to work the rich fisheries off the west and south-west coasts of Iceland. Sooner or later, during the outward passage, a few of them are likely to have been forced off their course by gales and driven away to the westward: or, again, they may have been blown off the fishing-grounds of the Vestmannaeyjar and elsewhere to the westward of the island and so driven to within sight of the Greenland coast;

which, in the region of Snaefellsnes, is only some three hundred miles distant. Similarly, the large English merchantmen are likely to have been driven off their course by adverse weather in the direction of Greenland; just as a number of Hansemen are recorded to have been driven there in the early years of the following century. After all, considering how often mariners from the days of Gunnbjörn Úlfsson to those of Thorstein Ólafsson had been forced westward to the Greenland coast, it would be quite inexplicable if nothing of the kind had from time to time occurred in the later fifteenth century, when so many trading and fishing voyages were made to Iceland.

It may well have been the lure of the stockfish that drew the English to Greenland, as it had earlier to Iceland. The immense demand for stockfish is revealed in the rapid exploitation of the Iceland fishery, in the number of ships employed in the traffic, and in the large cargoes brought back to England. At last, as the number of English ships venturing to Iceland continued to increase, there was not enough *skreið* to go round; and not a few of them had to leave with empty holds. In such circumstances the English would naturally be disposed to search for other sources of supply. Just as, a generation before, they had been forced by the situation across the North Sea to go further afield, they now sought out new fishing-grounds. Besides Iceland, there were the Faeroe Islands and Finmark; and also Greenland. As has already been said, they could scarcely fail to learn something of that country from the Icelanders, especially those in the western fjords; and they might even have come within sight of 'Greenland's icy mountains' through bad weather or other causes. If it was *skreið* they were after, there was enough and to spare in Greenland. Moreover, it is certain that Englishmen were fishing off the East Greenland coast in the latter half of the sixteenth century.[10] If in the wake of English fishermen had presently come merchantmen (as was certainly the pattern of events in Iceland), the imitation of European fashions in Greenland, and the apparent prosperity of Herjólfsnes in these later years, are easily explained.

The recurrent protests made by the King of Denmark concerning unauthorized English sailings to his overseas possessions may also have a bearing on the problem. Again and again reference is made in these complaints to Iceland, the Faeroe Islands, 'and our other dependencies'. Though in nearly every case Greenland is not specifically mentioned, but only the customary phrase 'and our other dependencies', nevertheless, in one document, there does exist a direct allusion to Greenland along with the other dependencies—'Island, Grønland, Ferrø, Hetland, Ørckenøer, eller andre Øer'.[11] The provenance of this particular document is admittedly not entirely certain; still, the leading authority on English traffic with Iceland in the Middle Ages, Björn Thorsteinsson, is of opinion that Greenland may be included among the lands from which intruding Englishmen had been 'warned off'.

It is even possible that certain products of Greenland are actually to be discovered in cargoes nominally ascribed to Iceland (or else the country of origin is discreetly concealed) in the English customs accounts. Both Carus Wilson and Thorsteinsson, indeed, have cited numerous instances of such concealment with respect to the Anglo-Icelandic traffic. During the era of baronial anarchy and general 'lack of governance', which is known in English history as the War of the Roses, a very large number of unlicensed and unrecorded voyages must have been made to Iceland; they may also have been made, for the same reasons, to Greenland.

In short, the sum of all the available evidence, literary and archaeological, seems to suggest that, throughout most of the fifteenth century, a number of ships did arrive in Greenland. The object, origin, and frequency of these sailings is unknown. But there was still, apparently, contact of some sort with the outside world.

Long after the date when official communications had ceased entirely, as the excavations at Herjólfsnes have revealed, certain articles of clothing produced in this almost forgotten outpost of Christendom continued to reflect the fashions of western Europe to a quite remarkable degree. It is to be noted that the garments themselves were not imported from overseas: they were the product of a local industry. Exactly how these fashions reached the Greenlanders may never be known; but the undeniable and astonishing fact remains that the very dress worn by contemporaries of Chaucer, Dante, and Petrarch was produced here in the old Norse colony 'at the world's end'. The discoveries include a distinctive type of pleated gown—one of them with a V-shaped neck—such as is depicted in the portraits of Christus and Pisanello: these were in vogue about the middle of the century. There are also a number of high conical caps, such as may be seen in the paintings of Dirk Bout, Hans Memling, and others of their school; allowing for the normal time-lag, they were probably worn in Greenland in the final quarter of the century. Besides these articles of clothing there are fragments of Rhenish stoneware and a knife from the episcopal kitchen at Gardar which tell the same story.[12]

Another remarkable fact which has emerged from these discoveries is that much of the clothing found at Herjólfsnes is not the dress of poor peasants, but of the relatively well-to-to.[13] From this it would seem that the harbour under Herjólfsnes, 'which is called Sandhaven, where merchants from Norway were wont to come'— in former days the first regular port of call for the Greenland *knörr*—continued to prosper even in the final phase of the medieval colony. It could hardly have done so without continued traffic with vessels from overseas or without some acceptable local product to offer in exchange for goods imported from Europe. In all probability that local product was *skreið*, or 'the comodius stokfysshe'.[14]

Finally, the archaeological evidence shows clearly that to the end of the chapter the inhabitants of Herjólfsnes, Sandnes, and other places in the colony remained faithful to Christian and European tradition. There is no sign of intermarriage with the Eskimo. 'The dead lie piously with folded arms, and not infrequently with a small wooden cross in the hands or on the breast; sometimes the cross was laid on top of the coffin lid.' In a number of cases the crosses are carved in obvious and unmistakable imitation of the Celtic cross. The inscriptions on some of them are a curious mixture of Latin and Old Norse, resulting in such hybrid invocations as *Jesus Kristi hjalpi*. The shortest inscription is one discovered at Brattahlid, 'Ingebjorg's grave'. In one corner of a coffin disinterred in the churchyard at Herjólfsnes was found a small stick inscribed with the runes: 'This woman, who was called Gudveg, was buried in the Greenland Sea'. The majority of the dead, however, were interred in their own frieze garments. Even the Bishop of Gardar was buried without a coffin.[15]

Though there is much controversy and uncertainty about the state of the climate about this time, it is safe to say that both the documentary and the archaeological evidence strongly suggest that climatic conditions in Greenland steadily deteriorated in the later Middle Ages. The Eskimo follow the seals, which frequent the edge of the ice, and this appears to indicate that the ice-edge, which was formerly high up in Baffin Bay, was gradually advancing southward. Mention has already been made of the significant change of course for shipping bound to Hvarf, necessitated by the presence of vast masses of ice in the Greenland Sea. In the Icelandic annals there are frequent references to drift-ice around the island.

The climatic deterioration recorded in the historical sources is borne out by the evidence of archaeology. From the Danish investigations of the last half-century it is certain that the East and West Settlements supported far larger flocks and herds than could possibly be raised there today. Farms have been discovered beside fjords that are at present blocked by ice. It has been shown, too, that in the vicinity of Herjólfsnes about the year 1400 the ground became permanently frozen. At the date when the bodies were interred it must have thawed in summer time to a considerable depth; for both the bodies and the garments in which they were shrouded were penetrated by the roots of plants. It is significant that the later burials lay a good deal nearer the surface. The result of these climatic changes was that the Greenlanders were faced with steadily declining food supplies. The land could no longer support the numerous livestock of former days.

Another significant contributory factor in the colony's decline was the gradual encroachment of the Eskimo. At the very time when the life of the Greenlander was becoming progressively harder, the pressure of the aboriginals began in earnest. Moving from hunting-ground to hunting-ground, the Eskimo came southward with the ice until, about the middle of the fourteenth century, they are recorded to have overrun the West Settlement; of which nothing more is heard.

No adequate explanation for the sudden extinction of the West Settlement is forthcoming. Not a word reached Europe of the fate of the inhabitants, though communication was still kept up, and kept up for many years to come, between Norway and the East Settlement. Various theories have been advanced to account for its disappearance. It has been argued that the inhabitants were wiped out by a pestilence. Several such epidemics had afflicted Iceland. Ships arriving from Europe may have brought the infection with them. There are indications of mass burials in the churchyard of the convent by the Siglufjörd. Another suggestion that has been made is that the pastures of the West Settlement were so ravaged by an invasion of the larva *Agrotis occulta* that they were fatally reduced in area and quality. Neither of these theories really accounts for the mysterious situation witnessed by Ívar Bárdarson. Ívar was instructed by the lawman at Gardar to proceed to the West Settlement and to drive the Skraelings out of the area. On their arrival Ívar and his men found that the place was deserted—there was not a soul to be seen, either Christian or heathen, only livestock running wild; of which they took away as many as their craft would hold, and then sailed back home.[16]

There are no grounds for supposing that there was inveterate hostility between the two races. The archaeological evidence, as well as legend and tradition, suggests that Norsemen and Eskimo might well live side by side in friendship. There are many examples of Norse influence upon Eskimo culture. The Eskimo in fact learned several useful arts from their European neighbours. In all probability the Eskimo language contains a few—though only very few—words of Norse origin.

Even the names of certain Norse chieftains were preserved in Eskimo legend: thus Yngvar and Ólaf became Ûngortoq and Ulâvik on the lips of their Eskimo neighbours.

On the other hand, the documentary evidence, Eskimo tradition, and the testimony of archaeology all point to a certain amount of frontier strife between the two peoples.[17] 'It needs no great effort of the imagination', writes Thorsteinsson, 'to see that little friendship is likely to have existed between the Greenland farmers and the Skraelings who, with their very limited understanding of the nature of property rights, must have gone for the sheep-flocks when their hunting failed— and at other times as well.'[18] Also it is rather too hastily assumed that, because the Eskimo of the present day is mild-mannered and peaceably inclined, his forbears of five or six hundred years ago were of much the same disposition. In 1355 the colony was alleged to be in such straits that a relief expedition was sent out from Norway;[19] and in 1379 the Eskimo are recorded to have made a damaging attack on the colonists, slaying some and leading off others into slavery.[20] It is to be presumed that the latter statement refers to some encounter which may have occurred in the distant Norðsetr.

According to one of the Eskimo legends taken down by Henry Rink about a century ago, an Eskimo in his kayak came up the fjord to the Norse settlement. Seeing a Norseman gathering shells on the beach, he slew him with a lance. The Norsemen retaliated with a murderous raid on the Eskimo, which was followed by an even more devastating assault on the part of the latter. The Norse chieftain, Ûngortok (Yngvar), escaped from his burning homestead with his little son in his arms. His Eskimo foe rapidly overhauled him. Yngvar in desperation hurled the child into a lake so that it might die uninjured. Relentlessly pursuing, the Eskimo finally caught up with Yngvar, put him to death, and then cut off his right arm. Another of these Eskimo legends tells of a bloody affray on the ice-bound fjord.[21]

Bearing in mind the likelihood, already discussed, that Greenland was visited, in the course of the fifteenth century, by ships from England, there is another possibility to be taken into account. That is, that the depredations for which the English had become notorious in Iceland were repeated in the far weaker and more vulnerable colony of Greenland.[22]

In the Danish and Icelandic archives there is recurrent record of English outrages in various parts of Iceland: of killings and woundings, of robberies, of burning and plundering of churches, and of the abduction, as in Finmark and the Faeroe Islands, of young boys and girls. In our own records reference is made to a number of English ports where, in the fifteenth and sixteenth centuries, Icelanders were said to be domiciled. It would appear that it was the younger folk who were chiefly sought after by the English marauders.[23]

The kidnapping apparently began in the early 1420s. It was laid down in one of the provisions of the treaty of 1432, which was concluded between the Kings of England and Denmark, that all prisoners carried off by the raiders were to be restored to their own country.[24] How far this part of the treaty was observed must remain an open question. Some, at least, of the Icelanders did not return home; for it is on record that a number of them took the oath of allegiance.

In the light of these developments it is worth examining again the letter written by Nicholas V in 1448 to Bishop Marcellus and considering whether the raiders mentioned there by the Pope should perhaps be taken as referring, not to the Eskimo, but to the English. It was stated in the letter that thirty years ago, from the neighbouring coasts of the heathen, 'the barbarians came with a fleet, attacked the inhabitants of Greenland most cruelly, and so devastated the mother-country and the holy buildings with fire and sword that there remain on that island [Greenland] no more than nine parish churches which are said to lie farthest away and which they could not attain because of the steep mountains'. The unhappy inhabitants of both sexes, and especially those who were thought fit and strong enough to bear the continual burden of slavery, they carried away captive to their own country. It was added that the majority of those taken prisoner had since returned to their own land and had in certain cases been able to repair their ruined homes; and that it was their earnest desire to restore and extend the service of God, since those who wished to attend divine service were obliged to travel for several days to reach such churches as had escaped the onslaught of the barbarians.[25]

Though the statement that the 'barbarians' carried off large numbers of Greenlanders into slavery is incredible when ascribed to the Eskimo, it is a different matter altogether if it is taken as referring to the English. It would be surprising indeed if some of the English rapscallions who had been ravaging Iceland and other dependencies of the Danish Crown had not sooner or later found their way to Greenland and committed similar depredations there. The list of misdeeds set out in the Pope's letter—the landing of armed marauders, the burning and looting, the destruction of churches, and the abduction of many of the inhabitants—does bear a significant resemblance to what is known to have occurred in Iceland.[26]

There are other indications besides that Greenland was ravaged by pirates. It is perhaps not without significance that no vestige of church vessels has been discovered in either the Norse or the Eskimo ruins. Also there are indications that Herjólfsnes, Gardar, and certain other churches were destroyed by fire (from the Eskimo Herjólfsnes received the name of Ikigait, 'the place destroyed by fire'). Lastly, a fragment of evidence which is well worth weighing is the oral tradition handed down among the Eskimo and related by a certain *angagok* (medicine man) to Niels Egede, son of Hans Egede, the evangelist of Greenland.[27]

According to this *angagok*, when his forefathers moved southward down the West Greenland coast they came at last to settle near the Norse inhabitants of the region, with whom they appear to have lived on friendly terms. Then there came a day when three small ships stood in to the land; and the crews robbed and killed a number of the Norsemen. The other Norsemen, however, beat off the attack, and the second ship sailed away, leaving the third in the hands of the Norsemen. But the following year there came many more ships; and the raiders overran the settlement, killing and looting and carrying off their cattle and other possessions, and finally sailed away. After this many of the Norsemen crowded into small open boats and made off down the coast; leaving a remnant of their number behind, whom the Eskimo promised to assist in the event of further attacks. But next year the raiders came again and the Eskimo fled in terror, taking a few of the Norwegian women and children with them up the fjord; 'leaving the others in the lurch', as the *angagok* expressively related. When in the autumn the Eskimo returned to the settlement they viewed with horror the havoc and destruction wrought by the raiders: the homesteads reduced to ashes, the farms laid waste, and all the movables carried off.

They thereupon took the Norwegian women and children and retreated high up the fjord, where they lived in peace for a number of years. The Eskimo took the Norwegian women to wife; 'five in number, with some children'.

Exactly when the East Settlement became extinct is a matter of conjecture, and will probably never be known. It is by no means impossible that a handful of survivors may have lived on after the turn of the century. We may even have to revise our views with respect to the story of 'Jón Grœnlendingr', which was for long rejected as apocryphal. According to this account, Jón, an Icelander, was sailing in 1540 or thereabouts from Hamburg to Iceland in a Hanse merchantman, which was driven off her course and eventually arrived in a fjord near the southernmost point of Greenland. On an island within this fjord were presently discovered 'many sheds and huts and stone houses for drying fish', and also the body of a man clad in frieze and seal-skin, with a hood on his head, lying face downwards on the ground. Beside him lay a sheaf-knife, bent, and much used and worn. Jón and his friends took the knife away with them 'as a keepsake'. The description of the dead man's attire certainly agrees with what we know of the habiliments of the fifteenth century Greenlanders.[28] The archaeological evidence also tends to support the theory that a few of the Greenlanders were living in the opening years of the sixteenth century.[29]

One thing, however, is certain. When in the autumn of 1585 John Davis sighted the east coast of Greenland and sailed down to Cape Farewell, and then, 'not being able to come neere the shoare by reason of the great quantitie of yce', northward again up the west coast, the Norse colony was extinct. The once thriving settlements lay ruined and desolate around silent churchyards wherein rested the dead of this forgotten outpost of European civilization; and Greenland was once more in the sole possession of the Eskimo.

22

The English voyages of discovery

IT WAS NO coincidence that the great voyages of discovery, at the conclusion of the fifteenth century and at the beginning of the sixteenth, took place when and where they did. The success and continuance of these ventures were dependent upon factors which were operative then and which were not operative in earlier periods of the world's history. Hitherto neither the ships nor the navigation of our ancestors had been equal to such undertakings. Towards the close of the fifteenth century, however, the situation was changing. In the forefront of a general and revolutionary advance at sea were the mariners of the Italian city states, and of Portugal and Spain.

Though England was relatively backward both in shipbuilding and in the knowledge and practice of scientific navigation, recent investigations have shown that the English share in the geographical discoveries of the age was, nevertheless, rather more important than was formerly supposed. The subject is still very obscure; but certain conclusions can safely be drawn. These English voyages of discovery were all confined to the North Atlantic, and originated in the activities of various merchants and mariners of Bristol.

In the fifteenth century Bristol, the second port in the kingdom, carried on an extensive commerce with Iceland, Gascony, Spain, Portugal, Ireland, and, in all probability, Madeira and the Canaries. Seeking further outlets for her trade, she also began to send ships to the Levant; and in the latter half of the century Bristol mariners are recorded to have been venturing far out in the Atlantic in search of a land called 'the Isle of Brasil', which was believed to lie somewhere to the west of Ireland.[1]

In 1480, according to the *Itinerarium* of William of Worcester, a relation of his named John Jay the Younger—a substantial merchant of Bristol, who had trading connections with both Iceland and the peninsula—dispatched an 80-ton ship, under one Thloyde, or Lloyd (described as 'the most skilful mariner in all England'), to search for this island. Thloyde left the Kingrode on 15 July (this was late in the year for a westbound passage), and on 18 September news reached Bristol that he had been compelled by stress of weather to abandon the quest and put back into an Irish port.[2]

Next year a second expedition, consisting of two small vessels named the *Trinity* and *George*, was sent out 'to serche & fynde a certain Isle called the Isle of Brasil'. It is significant that the two ships carried with them a large quantity of salt—which, it was explained, was not for merchandise, but 'for the reparacion and sustentacion of the said shippes'; from which it has been deduced that a major object of the enterprise may have been the discovery, or rediscovery, of new fishing-grounds.[3]

Nothing is known of the result of this expedition; nor is there any record of further attempts at Atlantic exploration during the same decade. That is by no means to say that no such attempts were made; there may well, in fact, have been others: after all, it was purely by chance that the two ventures mentioned above were actually set on record. But it would appear from a letter written in 1498 by the Spanish envoy in London, Pedro de Ayala, to Ferdinand and Isabella that, during the 1490s, the search for the distant Isle of Brasil continued. 'The people of Bristol', observed the envoy, 'have, for the last seven years, sent out every year two, three, or four light ships (caraveles), in search of the island of Brasil and the Seven Cities.'

What would appear to be a significant thread linking these developments was the close association of a group of Bristol merchants in the traffic with Iceland and Portugal, and also in the Brasil voyages of 1480 and 1481. The group included two men—Robert Thorne and Hugh Elyot—who at a later period were reported to have been 'the discoverers of the Newfound Landes'.[4]

These are facts which have been known to historians for a good many years. It was not, however, until fairly recently that a most interesting and important discovery by an American scholar engaged in research in the Spanish archives at Simancas cast a revealing light on these early Atlantic voyages out of Bristol. In 1956 Dr. L. A. Vigneras came upon a letter written in the winter of 1497–8 by a merchant named John Day, who had lately arrived in Andalusia from England, to a Spanish magnate styled 'the Grand Admiral', who is supposed to have been Christopher Columbus. The latter contains not only a long and detailed account of John Day's venture of 1497, but also refers briefly to a landfall made by some unknown vessel from Bristol en otros tiempos, 'in the past'.

'It is considered certain', Day informed 'the Grand Admiral', 'that the cape of the said land [from which Cabot had just returned] was found and discovered in the past by the men from Bristol who found "Brasil" as your Lordship well knows. It was called the Island of Brasil, and it is assumed and believed to be the mainland that the men from Bristol found.'[5]

Earlier in the century, therefore, it would appear that some mariners of Bristol had found their way to an unknown land on the other side of the ocean. When this occurred is a matter of conjecture. The difficulty is to know exactly what is signified by the phrase en otros tiempos, which may mean as long ago as thirty years or more; but may, on the other hand, signify a much shorter period. The Icelandic authority, Björn Thorsteinsson, with his intimate knowledge of the situation in Iceland and Greenland, favours the former interpretation, which is also supported by a number of Spanish scholars. It is to be remembered, too, that the writer of the letter (whether he translated it himself or not) was not a Spaniard but an Englishman. The precise meaning of the phrase, in fact, remains an open question.[6]

If the date of this discovery of Brasil is shrouded in doubt, it is to be emphasized that the same doubt extends to other areas of the problem. Thus, it can no longer be accepted as an absolute certainty that Columbus was the first to arrive in American waters in the age of the great discoveries. The date, 1492, is no longer the outstanding landmark in history that it was.

Nothing is known of the circumstances of this discovery. No record of it has ever been found, apart from the letter in question. It is not even known whether the mariners from Bristol were the first actually to sight the land; or whether they had learned of its existence from some other crew; or, again, acquired the knowledge

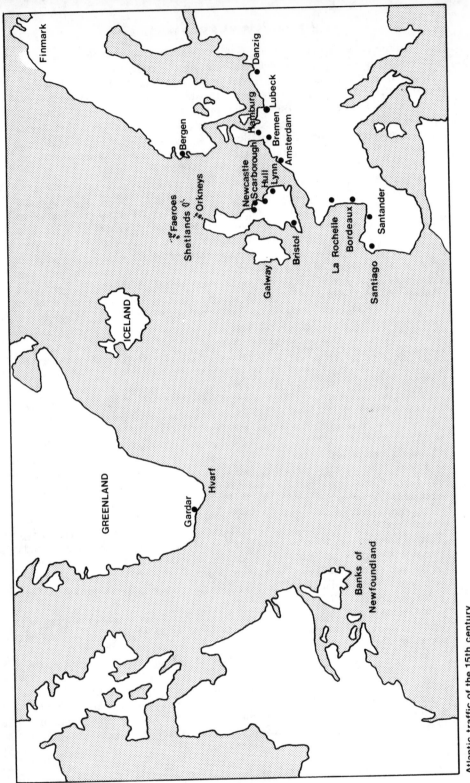

Atlantic traffic of the 15th century

on a visit to some Greenland haven. They may have made a planned voyage to the unknown shore; alternatively, as with sundry other Atlantic discoveries, the landfall may have been quite fortuitous. As was explained in the previous chapter, there is even a possibility that the English were sailing to Greenland as early as 1430; on such a passage they may sooner or later have been driven off their course, like Bjarni Herjólfsson, to within sight of the American coast. In any event the location of, and the course to be steered for, this distant land was seemingly no longer known; and numerous, though unavailing, attempts had been made over a term of years to rediscover it.[7]

In Bristol alone among English sea-ports at this time there was a strong objective interest in Atlantic exploration and discovery. There must surely have been some very powerful inducement to account for all the attempts that were made in later years to find the lost 'Isle of Brasil'. The expense, hardship, and hazard of such enterprises were considerable. What that inducement probably was is suggested by the reference already quoted to the quantity of salt included in the lading of the George and Trinity in 1481. For the preservation of cod, or stockfish, comparatively little salt was required; but the fish had to be taken ashore and thoroughly dried in the wind, as in Iceland, Greenland, and elsewhere.

It was the lure of dried cod, or skreið, which had brought the English to Iceland, Hálogaland, and Finmark in the fifteenth century; and also, judging from the Danish edict of 1431, possibly to Greenland as well. Bristol, Greenland, and 'the Isle of Brasil' form part, as it were, of a triangular problem whose solution is still very far to seek.

The rumour or report of some rich fishing-ground away to the westward would appear to be by far the most likely explanation of these Atlantic ventures in the 1480s and 1490s.[8] That there was land somewhere far out in the Western Ocean was almost certainly known to a number of merchants and mariners of Bristol. Apart from the great cod fishery off the shores of Newfoundland there was nothing in this region to lure them across the ocean.

The whole problem is further obscured by the factor of the weather and navigational conditions in this part of the North Atlantic; and also by the special circumstances of the Iceland and Greenland traffic: which, though forbidden to foreigners by law, was permitted in practice.

Navigation in this era was largely a matter of dead reckoning. Prolonged bad weather might throw out the mariner's most careful calculations. Surface currents, too, were an unknown quantity. In fact, the various unpredictable factors to be reckoned with were such that the successful outcome of an ocean passage made solely by D.R. might be attributed to chance rather than to skill[9] (especially if, as seems possible, they arrived off the North American coast by way of Iceland and Greenland). All this may explain why 'the Isle of Brasil', though it had been found at some former time, could not be rediscovered for so many years. Nevertheless, these men of Bristol apparently had some idea of where it lay, and of what parallel to seek to catch the easterly winds of spring.

The fact that trafficking with Iceland and Greenland was theoretically prohibited under heavy penalties explains the necessity for caution and circumspection in such ventures. Though the law might be, and frequently was, flouted with impunity, it nevertheless rendered both ship and lading liable to confiscation; and on numerous occasions, as we know, both ship and lading were summarily forfeited. That is the reason why so many of the English sailings to Iceland went unrecorded. If certain English vessels were also trafficking with parts of Greenland about this time, it was very much to their interest to conceal such ventures from the English and Danish officials.[10]

The English ventures were altogether independent of Latin influences and the motives which inspired Columbus, Cabot, and Vasco da Gama. Whatever it was that drew our mariners across the Western Ocean, it was not the dream of reaching the shore of Asia. It would seem likely that they were the first Europeans of their day to arrive off the eastern seaboard of continental America. It is even within the bounds of possibility that their discovery, in whatever way it may have come about, preceded that of Christopher Columbus by more than thirty years. In short, we can no longer be certain that it was the Bristol mariners who learned from Cabot, for it may have been the other way round, that is, Cabot who learned from the Bristol mariners—*that there was land to the north-west across the ocean.*

John Cabot, a Venetian by birth, in his younger days had personally taken part in the spice-trade; he had visited the great mart of Alexandria and had sailed down the Red Sea as far as Mecca. He was deeply versed in the travels of Marco Polo. He had pondered these matters long and earnestly before he evolved the project which was to engage him for the rest of his life. Cabot proposed, like Columbus, to reach the rich lands of Eastern Asia by a westward passage across the Atlantic. Only he did not consider that Columbus had gone far enough to the west; he did not believe that the shores that the latter had visited were anywhere near the empire of Cathay (China) or Cipango (Japan). He had endeavoured to interest the authorities in Portugal and Spain in his project, but he had had no success with either. At Bristol, however, he met with both keen interest and solid support. In February 1496 the King granted him letters patent.

'John Cabot's position in the enterprise may be seen more clearly since the publication of the John Day letter. The Bristol men had already found their Brasil', sums up J. A. Williamson. '. . . Cabot, seeking aid in Lisbon and Seville, where there were Bristol men to talk with, learned of it and decided that England must be his base. There was land in the west, well clear of the Spanish activities. This land could be a stage on the way to Cipango. . . . He would extend a fishing voyage into a new trade-route that would divert the richest of all trades to an English port.'[11]

From John Day's letter to 'the Grand Admiral' it would now seem that Cabot's voyage of 1497 was preceded by an earlier venture in 1496. 'Since your Lordship wants information relating to the first voyage, here is what happened: he went with one ship, his crew confused him [i.e. he had trouble with his men], he was short of supplies and ran into bad weather, and he decided to turn back.' On his first voyage, in fact, Cabot was no more successful than Thloyde had been in 1480.[12]

The following year he made another attempt. In the little *Matthew*, manned by a crew of twenty or so, including two or more merchants of Bristol, he sailed from England towards the end of May. He shaped a course round the south coast of Ireland, and then stood north to a parallel on which they continued during the greater part of their westward passage. The weather favoured them. It appears that during most of the outward passage the wind was E.N.E. and the sea smooth; though three days before sighting land they ran into bad weather, which, however, only lasted for a day. Day observes that before they came to land they passed through a

region of the ocean where the variation of the compass amounted to two points or
$22\frac{1}{2}$ degrees westward.[13] No mention is made of astrolobe or quadrant: but they
seem to have had some idea of relative latitude, and of the proper course to be
shaped for the distant 'isle'.

Cabot made his landfall at dawn in the last week of June—probably on the
twenty-fourth. Later in the day he went ashore with a few companions and erected
a crucifix and the banners of Henry VII, the Pope, and St. Mark of Venice. Beyond
the shore were tall trees 'of the kind of which masts are made' and wide pastures. It
was fine summer weather and very warm. Though in their brief excursion they
came upon no inhabitants, there were unmistakable signs that such existed.
According to one account, 'they found a trail that went inland, they saw a site where
a fire had been made, they saw manure of animals which they thought to be farm
animals, and they saw a stick half a yard long pierced at both ends, carved and
painted with brazil.' Another account adds that they found certain snares which
were spread to take game and something which looked like a needle for making
nets; they also discovered trees which had been felled. The party did not stay long
ashore. Though the groves around them appeared wholly deserted, they evidently
went in fear of sudden attack. They were so few, relates John Day, that 'he did not
dare advance inland beyond the shooting distance of a crossbow, and after taking in
fresh water he returned to his ship'.[14]

For a month or so they stood eastward along the coast, without landing, and then
returned to a cape of the mainland which they reckoned to be in approximately the
same latitude as Dursey Head in Ireland. All along the coast, John Day observes,
they found many fish like those which in Iceland are dried in the open and sold in
England and other countries, and are called in English 'stockfish'. Thus following
the coast 'they saw two forms running on land one after the other, but they could
not tell if they were human beings or animals; and it seemed to them that there were
fields where they thought might also be villages, and they saw a forest whose foliage
looked beautiful'.[15]

They sailed from the above-mentioned cape of the mainland with a fair wind and
(according to John Day) reached Europe in fifteen days. They made their landfall,
however, in Brittany. The error was not Cabot's but that of his companions, who
insisted that he had been steering too far north.[16] From Brittany they sailed to
Bristol, and Cabot thereupon journeyed to London to report to the King.

Cabot's reception at the court of Henry VII was all that he could have desired. He
had become the hero of the hour. Despite the opinions of the geographers, it was
widely believed that on the other side of the ocean, beyond 'the Isle of Brasil', lay
the country of the Grand Khan and the fabulously wealthy island of Cipango.
There was excited talk of colonization and a rich new trade-route. Rewards were
showered, not only on the leader of the expedition, but also on his followers. In vain
Ayala protested that the land discovered by Cabot lay within the sphere of Spanish
sovereignty. Henry VII would not have it.[17]

What particularly impressed Cabot's English companions was the incredibly
prolific cod fishery they had discovered—or rediscovered. They asserted that the sea
along the coasts they had visited was swarming with fish; of which they said they
could now bring home such vast quantities that the kingdom would have no
further need of Iceland. But Cabot's mind was set upon greater things than
stockfish. It was his intention to work eastward along the land which he had

discovered last summer until he came to the island of Cipango: from which island he firmly believed all the spices of the Orient were derived. By this means he hoped to make London a more important mart for spices than Alexandria itself.[18]

'The King', declared Lorenzo Pasqualigo, a Venetian merchant in London,

has promised him for the spring ten armed ships as he [Cabot] desires and has given him all the prisoners to be sent away, so that they may go with him, as he has requested, and has given him money that he may have a good time until then, and he is with his Venetian wife and his sons at Bristol. His name is Zuam Talbot [John Cabot] and he is called the Great Admiral and vast honour is paid to him and he goes dressed in silk, and these English run after him like mad, and indeed he can enlist as many of them as he pleases, and a number of our rogues as well.[19]

In the spring of 1498 another expedition, led by John Cabot, was fitted out by the merchants of Bristol and London, with some assistance from Henry VII (four ships were equipped by the former, and one by the latter). The expedition left Bristol in May, and appears to have been dispersed in a gale. The ship in which Cabot sailed put into an Irish port and thereafter stood to the westward alone. That was the last that was ever heard of John Cabot.[20]

What became of the four other ships is unknown. It is very unlikely that they were all lost, and there are certain indications that one or more of them may have stood southward for a considerable distance down the American coast. No positive knowledge exists of any discoveries made by this expedition. It is significant, however, that after the voyage of 1498 there was no more talk in England of the land across the ocean as forming part of Asia. The references are always to 'the New Land' or the 'New Found Land'.[21]

The men of Bristol, in conjunction with certain Portuguese colonists of the Azores, followed up this venture with further voyages to the north-west during the years 1501 to 1506, and possibly still later; and towards the end of 1502 a syndicate which became known as 'The Company Adventurers to the New Found Land' was formed—the first association for overseas expansion in our history. An entry in the Privy Purse accounts records that, on 30 September 1502, a sum of £20 was paid out to 'the merchants of Bristol that have been in the New found Land', while two of the Portuguese were awarded pensions of £10 a year each for having acted as 'captains' to the 'New Found Land'. Sebastian Cabot, the son of John Cabot, who made two voyages to the north-west, in 1504 and in 1508–9, also received an award for 'the fyndynge of the newe founde landes'. In 1504 two ships returned to Bristol with cargoes of salted fish and fish livers from the 'Newe Found Iland'. These ventures progressively increased the common stock of knowledge of the North American littoral. There are even indications that about this time some kind of settlement may have been established on the shores of Newfoundland or Nova Scotia: but of this there is no positive evidence.[22]

What may perhaps have been in past decades the private and particular knowledge of a limited number of mariners now became public property. The rich cod fishery off the coasts of Newfoundland began to be worked by fishermen from various parts of western Europe. In the earliest years of the sixteenth century the Portuguese sailed there in such numbers that by 1506 a special tax was imposed on Newfoundland cod imported to their country. Shortly afterwards they were joined by Normans and Bretons from France, and also by Basques from Spain. In 1508 four ships were fitted out in Rouen for a voyage to 'the New Land'. An English explorer, John Rut, encountered no less than eleven French fishing craft in the haven of St. John's in the summer of 1527.[23]

Owing to lack of evidence it is difficult, if not impossible, to assess the English participation in the new fishery. A considerable area of uncertainty persists during these later years. When towards the end of the sixteenth century Richard Hakluyt was engaged in compiling his famous collection of English voyages, he had occasion to lament 'the great negligence of the writers of those times, who should have used more care in preserving of the memories of the worthy acts of our nation'. It is at any rate clear that there were a number of English ventures of which all direct record has been lost, and are known to us only by implication. Thus in the *New Interlude*, written in 1519 by John Rastell, brother-in-law of Sir Thomas More (which contains the first English account of the New World), reference is made to this traffic in general terms.

> This see is called the great Oceyan
> So great it is that never man
> Coude tell it sith the worlde began
> Tyll nowe within this xx yere
> Westwarde be founde new landes
> That we neuer harde tell of before this
> By wrytynge nor other meanys
> Yet many nowe haue been there . . .

Also in a letter from the Company of Drapers to Henry VIII and Wolsey in 1521 mention was made of the English-born 'maisters & mariners' who had experience of the navigation to the New Land, 'aswele in knowledge of the land, the due course of the seey, thiderward & homeward, as in knowledge of the havenes, roodes, poortes, crekes, dayngers & sholdes there uppon that coste'. It is known, moreover, that the mariners of Bristol forsook the old trade-route to Iceland in favour of Newfoundland, and that other west-coast ports followed their example; though it would appear that the English contingent in the cosmopolitan Newfoundland fleet was by no means large until much later in the century.[24]

Year after year these fishermen would sail for Newfoundland with the easterly winds of spring. Judging from the numbers that presently engaged in the traffic, the navigation to Newfoundland seems to have presented few difficulties. It was found by experience that, during this time of the year, 'one wind suffiseth to make the passage, where as most of your other voyages of like length, are subject to 3. or 4. winds'.[25] Among the Portuguese and French, some fished on the banks; but the majority foregathered in St. John's, from which haven they spread out, north and south, along the coast to engage in the shore fishery, returning to the rendezvous at St. John's in the autumn. Their activities are commemorated to this day in such place-names as Portuguese Cove, Spaniards' Bay, Biscayan Cove, Frenchmen's Cove, English Harbour, and so on. The frequency of English names on the east coast of Newfoundland bears testimony to the early occupation of these havens by the English, who from the outset concentrated exclusively on the shore fishery.

These sturdy fishing craft represented the first regular traffic on the North Atlantic run. Henceforward, at first a mere trickle, and later a steady flow of shipping, made the long transatlantic venture. No longer was it a question of sporadic and unsuccessful voyages in search of mythical islands. Their destination was the richest, most prolific fishing-ground in the Western Ocean. The course for the distant shore off which the fishery lay was *known*, and that knowledge was never again to be lost. This marked an era in the history of the sea. The ocean had been bridged. As time went on the traffic increased, and it went on increasing. And all this began when some mariners of Bristol happened to light upon what became known to them as 'the Isle of Brasil'; they lost it, and then they found it again; and thus consummated the work begun in past centuries by Cormac ua Laith, Bjarni Herjólfsson, Eirík Raudi and his son Leif, Thloyde, and many an unknown mariner besides.

To what extent these developments were influenced by the new methods of navigation represented by the astrolabe and quadrant is unknown and likely to remain so. Down to the present time no evidence has come to light that any such instruments were used in these ventures. Their adoption was by no means universal even half a century later, when William Bourne wrote the *Regiment of the Sea*. (Bourne declared that the more conservative mariners, 'them that wer auncient masters of shippes', would deride and mock those who had recourse to the new methods.) It was in fact only by slow and gradual stages that the knowledge and practice of scientific navigation became a significant factor in the transatlantic traffic.

Cabot's dream of a westward passage to eastern Asia by way of 'the Isle of Brasil' and the diversion of the spice trade to Bristol had after all come to nothing. The fishery remained. As a school for deep-water sailors, the Newfoundland traffic was destined to be more important, even, than the trade to Iceland had been. 'It is the longe voyaiges', as Hakluyt remarked, '. . . that harden seamen and open unto them the secretes of navigation.' Successive generations of English mariners were to serve their apprenticeship in this rigorous school, which annually employed increasing numbers of deep-sea fishermen, drawn chiefly from our western ports. In the words of Sir Walter Raleigh: 'The Newfoundland fishery was the mainstay and support of the western counties.'

The glamour and drama of the achievement of Christopher Columbus and Vasco de Gama and Amerigo Vespucci have tended to obscure the significance of these relatively little-known developments across the ocean. Not only then, but for long afterwards, popular attention was in fact almost wholly centred on the spectacle of the treasure of the New World and of the Orient flowing into the coffers of Spain and Portugal; of exotic empires toppling before the superior armament and organization of their European conquerors. What was happening far out in the western Atlantic went virtually unnoticed. A few lonely English craft tossing in the waste of waters beyond the limits of the known world made little impact on contemporary thought, and in a number of cases even the record has been lost.

The navigation of the North Atlantic was a cardinal factor in the development of English maritime enterprise. A substantial number of seafaring communities all round the coasts of England had by now acquired a valuable stock of ocean experience. Already the foundations of that future prowess at sea, which was to earn the unstinted respect of those shrewd observers of maritime affairs, the Venetians, were being well and truly laid. Before the end of the sixteenth century John Davis could justly claim that in seamanship 'wee are not to be matched by any nation of the earth'.

It can scarcely be over-emphasized that the ocean navigation of the English in this era was a vital link in the causal chain of maritime history. Over and above the immediate stimulus to English seamanship already noted, the Newfoundland fishery was destined to become a major element of our future naval strength. Furthermore, by these efforts in discovery and their sequel the first steps had been taken towards the eventual domination of the North American continent by the English-speaking peoples.

Notes

1. Curachmen of the west

1. Private information: Dr. James O'Brien of Kilronan, Aran Islands, and others.

2. Antiquity of the skin-covered craft

1. J. G. D. Clark, *Prehistoric Europe* (London, 1952), p. 283.
2. G. Gjessing, *Fangstfolk* (Oslo, 1941), p. 91.
3. Brøgger and Shetelig, *The Viking Ships* (Oslo, 1953), p. 47. The late Professor A. W. Brøgger was one of the few—the very few—scholars who would go to professional seamen for advice on matters of seamanship and navigation.
4. B. Faerøvik, 'Farkoster i Norderlandi før Vikingetidi', *Bergens Sjøfartsmuseums Årshefte* (1937), pp. 12–14; Gjessing, *Yngre Steinalder i Nord-Norge* (Oslo, 1942), pp. 493–4; Clark, pp. 283, 289.
5. There is an interesting tribute to the seaworthiness of the skin-covered *umiak* of the Eskimo in Vilhjálmur Stefánsson's *Ultima Thule* (1942), p. 39. Stefánsson quotes the considered opinion of Captain E. D. Jones, Commandant of the U.S. Coastguard Academy, New London, Conn., who states that 'the *umiak* . . . is perfectly capable . . . of remaining afloat in almost any weather'. It should be added, that the experiences of the First and Second World Wars have considerably broadened our ideas as to what is possible in the matter of open-boat voyages.
6. *Fangstfolk*, pp. 86–7; Clark, op. cit., pp. 39, 254, 283; Brøgger and Shetelig, op. cit., p. 14.
7. *Aeneid*, vi, 413–14.
8. 'Sometimes about this Country, are seen these men they call *Finn-men*. In the year 1682, one was seen in his little boat at the South end of the Isle of *Eda*, most of the people of the Isle flock'd to see him, and when they adventur'd to put out in a Boat with Men to see if they could apprehend him, he presently fled away most swiftly. And in the year 1684, another was seen from *Westra*': James Wallace, *An Account of the Islands of Orkney* (Edinburgh, 1700), p. 60. See also Paul Gaffarel, *Histoire de la découverte de l'Amérique* (Paris, 1892), pp. 167–72.
9. Brøgger and Shetelig, op. cit., p. 14.
10. E. G. Bowen, *Britain and the Western Seaways* (London, 1972), p. 25. See also ibid., pp. 36–9.

3. The curach era

1. *Festus Avienus*, ed. A. Berthelot (Paris, 1934), p. 34.
2. *De bello civile*, I, cap. iv.
3. *Historia Naturalis*, iv, 104; ibid., 206.
4. *Pharsalia*, iv, 130–5.

5. *Polyhistor* (ed. 1689), cap. xxii, p. 31.

6. *Curach* is the Old Irish word for the hide-covered craft.

7. *Tripartite Life of St. Patrick*, ed. W. Stokes (London, 1887), ii, 446–7. It is to be noted that both in the *Tripartite Life of St. Patrick* and in the *Immram curaig Maíle Dúin* the curach is also called a *noie*. See Stokes, op. cit.; *R.C.*, vol. 9, pp. 459–60. At the present time the sea-going curach is called a *naohóg* in Co. Kerry. Cf. Roderick O'Flaherty, *Ogygia* (London, 1685), p. 250: 'Hibernicé *Corach* vel *Noemhog* vocantur'.

8. Sidonius, *Poems and Letters*, ed. W. B. Anderson (London, 1936), p. 150.

9. S. S. Frere, *Britannia* (London, 1974 ed.), pp. 350, 362.

10. Gildas, *De excidio et conquestu Britanniae*, ed. W. Winterbottom (Chichester, 1978), p. 19.

11. Ibid., pp. 14, 16, 19.

12. Claudian, *De consulatu Stilichonis*, ii, 250–5; *Tripartite Life of St. Patrick*, ii, 357; *The Life of St. Columba*, ed. W. Reeves (Dublin, 1857), p. 176.

13. A. de la Borderie, 'Saint Efflam', *Annales de Bretagne*, vol. 7 (Rennes, 1896), pp. 288–9, 302–3 and also *Histoire de Bretagne* (Rennes, 1896), p. 361; N. K. Chadwick, *Early Brittany* (London, 1969), p. 199.

14. Reeves, op. cit., pp. 168–9, 178. Cf. W. Stokes, 'The Voyage of Mael Duin', *R.C.*, vol. 9, p. 472: 'Luid cloch dib íarom ina curach co rotreagdastair sciath Mael dúin ocus ndechniadh [is] an drumluirg in curaig'. The modern Irish rowing-curach has, of course, no keel. One or two leeboards used to be employed by certain of the larger Dingle curachs, enabling them to carry a bigger sail than would otherwise be possible with safety, and, moreover, serving to reduce leeway: see James Hornell, *Water Transport* (London, 1946), p. 145. Hornell has argued that there is no real evidence of hide-covered wicker vessels fitted with keels, and has cast doubt on the statements of Julius Caesar and 'Captain Tho. Phillips' and others; but he appears to have overlooked the testimony of the *Vita S. Columbae* and *Immram curaig Maíle Dúin*. Cf. James Hornell, *British Coracles and Irish Curraghs* (London, 1938), Sect. III, p. 35.

15. Reeves, op. cit., p. 120.

16. Ibid., pp. 106, 176.

17. Ibid., p. 176.

18. Ibid., pp. 64, 77; *R.C.*, vol. 9, p. 494. The small curach was frequently used for penitential purposes. Cf. *Tripartite Life of St. Patrick*, ii, 288; Reeves, op. cit., p. 77.

19. *Navigatio Sancti Brendani Abbatis*, ed. C. Selmer (Notre Dame, Ind., 1959), pp. 10–11; D. Dumville, 'Immrama', *Ériu*, vol. 27 (Dublin, 1977), p. 89. In a Middle Welsh poem by Maredudd ap Rhys is a reference to tallow-cake being used to work the hide. See *British Coracles and Irish Curraghs*, Sect. III, p. 13.

20. *R.C.*, vol. 10, pp. 86–7.

21. *Bethada Náem nÉrenn*, ed. C. Plummer (Oxford, 1922), i, 65; *Vitae Sanctorum Hiberniae ex Codice olim Salmanticensi nunc Bruxellensi*, ed. W. W. Heist (Brussels, 1965), p. 25.

22. *De bello gallico*, III, cap. xiii.

23. Plummer, op. cit., i, 64, 71–3, 77; *R.C.*, vol. 10, pp. 60–1; Reeves, op. cit., pp. 155–6; Lb., p. 43; Dicuil, *De mensura orbis terrae*, ed. J. J. Tierney (Dublin, 1967), p. 76.

24. *Mabinogion*, ed. and trans. Gwyn and Thomas Jones (London, 1949), p. 32.

25. *Chronicles of the Picts and Scots*, ed. Skene (Edinburgh, 1867), p. 69.

26. *Annals of Ulster*, ed. W. Hennessy (Dublin, 1887), p. 205.

27. Hennessy, op. cit., p. 327.

28. Jean Young, 'A Note on the Norse Occupation of Ireland', *History*, N.S., vol. 35, p. 11; Alexander Bugge, *Norse Loan Words in Irish* (Christiania, 1912), *passim*; *R.C.*, vol. 11, pp. 493–5; vol. 12, pp. 460–2.

29. See *The Lives of S. Ninian and S. Kentigern*, ed. A. P. Forbes (Edinburgh, 1874), cap. x. After describing the small one-hide curach the writer, Ailred of Rievaulx, comments:

'Forte tunc temporis eodem modo naves immense magnitudinis parabantur.' Highland folklore may also preserve the memory of the old sailing-curach: see J. F. Campbell, *Popular Tales of the West Highlands* (Edinburgh, 1890), ii, 481.

30. Giraldus Cambrensis, *Topographia Hibernica*, ed. J. F. Dimock (London, 1867), Dist. III, cap. xxvi.

31. Alexander Carmichael, *Carmina Gadelica* (Edinburgh, 1928), ii, 228, 265. On the other hand, the *curachan* referred to by J. F. Campbell, op. cit., ii, 481, appears to have been a true sea-going craft: 'Chuir e'n curachan a mach, 's schuidh e innte 's cha d'rinn e stad na fois gus an deachaidh e air tir an Eirinn.'

32. *Water Transport*, p. 142.

33. *Ogygia*, p. 250.

34. *British Coracles and Irish Curraghs*, p. 10.

35. *Historicae Catholicae Iberniae Compendium* (Lisbon, 1621), III, Lib. vii, cap. ix, 190–1; Standish Hayes O'Grady, *Pacata Hibernia* (Dublin, 1896), ii, 283; *Water Transport*, p. 136.

36. Bibliotheca Pepysiana: *Sea Manuscripts*, 2934, f.41. See also R. Morton Nance, 'Wicker Vessels', *M.M.*, vol. 8, pp. 199–205; *Journ. Roy. Soc. Ant. Ireland*, Ser. VI, vol. 16, ii, p. 119; *British Coracles and Irish Curraghs*, Ser. III, pp. 35–7; Bowen, op. cit., p. 39.

37. *Ulster Journal of Archaeology*, vol. I (1853), p. 32.

38. C. H. Hartshorne, *Early Reminiscences of the Great Island of Aran* (Dublin, 1853), pp. 292–3. It was imperative in a gale to keep the curach head on to the wind (in the vernacular, 'to split the wind'). 'Let her fall off two inches', it was sometimes said, 'and she cannot be managed.' The greatest skill was called for in the management of the oars; for an untimely jerk might break the oar or thole-pin with calamitous consequences. Finally, so delicately balanced was this type of craft that it was always necessary to move about in her with the utmost caution; and at other times to keep as still as possible, and even 'to keep the tongue in the middle of the mouth' (personal experience; private information).

39. J. M. Synge, *The Aran Islands* (London, 1907), pp. 97–8.

40. Private information: Dr. James O'Brien, of Kilronan, Aran Islands.

4. Early ventures in the Atlantic

1. See W. R. Kermack, *Historical Geography of Scotland* (2nd ed. Edinburgh, 1926); O. G. S. Crawford, 'The Western Seaways', *Custom is King*, ed. Dudley Buxton (London, 1936), pp. 195, 200.

2. Einar Ól. Sveinsson, *Landnám i Skaftafellsþingi* (Reykjavik, 1948), pp. 10 sq.; W. F. Thrall, 'Clerical Sea Pilgrimages and the "Imrama"', *The Manly Anniversary Studies* (Ottawa, 1923); C. Plummer, *Vitae Sanctorum Hiberniae* I (Oxford, 1910), p. xli, n 2; H. Zimmer, *Zeitschrift für deutsches Alterthum*, vol. 23, p. 298; Heist, op. cit., p. 17 *et passim*.

3. For the visit of St. Brendan and Cormac ua Liatháin to St. Columba on the island of Hinba (probably Eilean na Naoimh in the Garvellach Islands), see Reeves, *The Life of St. Columba*, pp. 219–20.

4. Thrall, op. cit., pp. 278–83. See also N. K. Chadwick, *The Age of the Saints in the early Celtic Church* (London, 1961), pp. 72 sq.

5. Frere, *Britannia*, p. 2. See also L. Bieler, Ireland: *Harbinger of the Middle Ages* (London, 1963), p. 10.

6. Reeves, op. cit., p. 7. 'A striking feature of Celtic asceticism was the habit of retiring to islands for rest or to find a hermitage' (Thrall, op. cit., p. 277). Cf. Reeves, op. cit., pp. 86–7, 127, 135, 165, 197–8, 219–23, 288. The same tendency is reflected in the *immrama*. See *R.C.*, vol. 10, pp. 80–1, *et passim*. See also E. G. Bowen, *Britain and the Western Seaways* (London, 1972), p. 76.

7. Thrall, op. cit., p. 77 n 2; J. F. Kenney, *Sources for the Early History of Ireland* (New York, 1929), pp. 487–9; *Christian Art in Ancient Ireland*, ed. A. Mahr (1941), pp. 77–8; *The Age of the Saints in the Early Celtic Church*, pp. 95 sq.

8. Reeves, op. cit., pp. 30, 168–9; J. T. Fowler, *The Life of St. Columba* (Oxford, 1920), p. 157.

9. Reeves, op. cit., pp. 168 sq. Cf. *R.C.*, vol. 9, p. 468; vol. 10, p. 60.

10. Reeves, op. cit., pp. 49–50.

11. Frere, op. cit., p. 379; Bowen, op. cit., p. 70; Gildas, *De excidio et conquestu Britanniae*, ed. M. Winterbottom (Chichester, 1978), p. 2.

12. Standish Hayes O'Grady, *Silva Gadelica* (London, 1892), p. 352.

13. *Chronicles of the Picts and Scots*, ed. W. F. Skene (Edinburgh, 1867), *passim*; *R.C.*, vol. 35, p. 272; vol. 39, p. 304. See also Kenneth Jackson, *Language and History in Early Britain* (Edinburgh, 1953), pp. 27; John Bannerman, *Studies in the history of Dalriada* (Edinburgh, 1974), pp. 148–54.

14. Kermack, op. cit., p. 12; Crawford, op. cit., pp. 192, 195, 200; Dillon and N. K. Chadwick, *The Celtic Realms* (London, 1967), p. 70.

15. The coasts of South Wales, Cornwall, and Brittany present somewhat similar conditions. See Bowen, op. cit., p. 37.

16. H. M. Chadwick, *Early Scotland* (Cambridge, 1949), p. 152; Dobbs, 'Aedan mac Gabrain', *S.G.S.*, vol. 7, pp. 90–3; Dillon and Chadwick, op. cit., pp. 75–8, 81.

17. Skene, op. cit., pp. 69, 73, 345, 353; Dillon and Chadwick, op. cit., pp. 76–8.

18. H. M. Chadwick, op. cit., pp. 124–5; Isabel Henderson, *The Picts* (London, 1967), pp. 38–40, 47–51; Dillon and Chadwick, op. cit., pp. 74–81, 182; Bannerman, op. cit., pp. 79, 85–6, 148–54; Gildas, *De excidio et conquestu Britanniae*, ed. M. Winterbottom (Chichester, 1978), cap. 14.

19. H. M. Chadwick, op. cit., pp. 46–8; Bannerman, op. cit., p. 153; M. Miller, 'Bede's use of Gildas', *E.H.R.*, vol. 90 (1975), p. 244.

20. A. R. Lewis, *The Northern Seas* (Princeton, 1958), pp. 142–3.

21. Reeves, op. cit., pp. 167–8.

22. *Reports of the Royal Commission on Ancient and Historical Monuments of Scotland: Orkneys and Shetlands* (1946), i, 41; ii, 23, 184, 321, 337; iii, 44, 74; Sveinsson, op. cit., pp. 21–2, 35; *Monumenta historica Norvegiae*, ed. G. Storm (Christiania, 1880), p. 89. It was in the Orkneys and Shetlands, as well as in the Faeroe Islands, that the Norsemen first came into contact with the Irish monks (known to them as *papar*) whom they were later to encounter in Iceland.

23. *De mensura orbis terrae*, ed. Tierney, pp. 74, 76.

24. O. Rygh, *Norske gaardnavne* (Oslo, 1924), i, 265; *Stedsnavn*, ed. Magnus Olsen (Oslo, 1939), p. 53; Sveinsson, op. cit., pp. 16 sq., 23.

25. A. W. Brøgger, *Ancient Emigrants* (Oxford, 1929), p. 55.

5. The first discovery of Iceland

1. Alexander Bugge, *Den norske sjøfarts historie* (Oslo, 1923), p. 35. Professor Hawkes, relying upon the somewhat tenuous observations of Strabo and Pliny, has recently resurrected the old theory that Iceland was the land of 'Thule' visited by Pytheas and others. However, he adduces no sort of corroborative evidence and does not in fact dispose of the arguments which have always militated against the theory. See C. F. C. Hawkes, *Pytheas: Europe and the Greek Explorers* (Oxford, 1977), pp. 34–7.

2. Kristján Eldjárn, 'Fund af Romerske Mønter pa Island', *Nordisk Numismatisk Årsskrift* (Oslo, 1949), p. 8 *et passim*.

3. Haakon Shetelig, 'Roman Coins found in Iceland', *Antiquity* (Gloucester, 1951), vol. 23, pp. 161–3.

4. Cf. Björn Thorsteinsson, 'Íslands- og Grænlandssiglingar Englendinga á 15. öld og fundur Norður-Ameríku', *Saga* (Reykjavik, 1965), pp. 10–1.

5. Whitley Stokes, 'The Voyage of Mael Duin', *R.C.*, vol. 10, p. 53.

6. See *R.C.*, vol. 9, p. 463: 'Lotar íarom isan ocian mór nemfoircendach'.

7. Ibid., pp. 492–3: 'insi n'áird slíabdai lán d'énaib'.

8. Ibid., pp. 480–1.

9. R.C., vol. 10, p. 61.

10. See Zimmer in *Zeitschrift für deutsches Altertum*, vol. 30, o. 289; Plummer, *Vitae Sanctorum Hiberniae*, I, p. xli n. 2; Thrall, 'Clerical Sea Pilgrimages and the "Imrama"', pp. 278 n. 1, 283 n. 2; Myles Dillon, *Early Irish Literature* (Chicago, 1948), pp. 124–5.

11. See Plummer, *Bethada Náem nÉrenn*, ii, 70. In the same source there is a vivid account of a volcanic eruption, which is worth quoting in full: 'Another day as Brendan was voyaging on the ocean he saw a great hellish mountain which appeared full of clouds about its summit. And the wind carried them with irresistible force to the shore of the island whereon the mountain was, so that the boat was close to land. And the brink of the island was of an appalling height, so that they could scarcely see [the top of] it, and it appeared full of firebrands and red sparks, and was as steep as a wall. . . . And after this the wind swept them away, and they drove towards the south. And they looked back on the island, and saw it all on fire, belching out its flame into the air, and swallowing it again, so that the mountain seemed all one ball of fire' (p. 64). Turville-Petre, commenting on these lines, has observed that the passage 'would not be an over-fanciful description of Iceland during a volcanic eruption. It would apply best to the south or south-east of the island, to the precipitous Mýradalsjökull or to Öroefajökull' (G. E. O. Turville-Petre, *The Heroic Age of Scandinavia* (London, 1951), pp. 95–5). But it is to be remembered that this passage *may* have been introduced into the Brendan legend in consequence of the later experiences of Irish monks or Vikings.

12. H. L. Sawatzky and W. H. Lehr, 'The Arctic Mirage and the Early North Atlantic', *Science*, vol. 192 (New York, 1976), pp. 1300–4. The manifestation of the Arctic mirage is vividly described in the reminiscences of one of the officers serving in the 10th Cruiser Squadron during the First World War, H. H. Smith, *A Yellow Admiral Remembers* (London, 1932), p. 290: 'At one moment there is an absolutely clear horizon and to all appearances there is nothing beyond that horizon for miles and miles and miles. Then suddenly a change takes place in the atmosphere and a vast mass of snow-capped mountains, sparkling glaciers and deep valleys stand out against the sky, shortly afterwards to disappear behind the same horizon.' There was another and even more remarkable manifestation of the mirage, recorded by the master of an American schooner in 1939, when the Snaefellsnesjökull and other well-known landmarks in Iceland were sighted at a distance of more than 300 miles. See V. Stefánsson, *Ultima Thule* (New York, 1940), p. 51. See also Hydrographic Dept., Admiralty, *Arctic Pilot*, ii, xxxvii–xxxviii.

13. J. A. Beckett, *Island Adventure* (London, 1934), p. 26; J. Hornell, 'The Role of Birds in Early Navigation', *Antiquity*, vol. 20 (Gloucester, 1946), p. 146.

14. Dicuil, *De mensura orbis terrae*, ed. Tierney, p. 74; *I.F.* Ii, 5, 31–2. See also *Monumenta historica Norvegiae*, ed. Storm, pp. 8, 92.

15. It is as well to remember that even in the 'curach era' wooden vessels were also used by the Irish. The monks of Iona owned craft built of timber as well as curachs. The number of different types of craft listed by Adomnan in the *Vita S. Columbae* is surprisingly large. Reference is made to the *alnus, caupellus, cymba, cymbula, naves longae dolatae, navicula, navis oneraria*, and *scaphae*. Adomnán records the arrival, off the coast of Kintyre, of a Gallic bark, *barca* (Reeves, *The Life of St. Columba*, pp. 176 n. 363).

16. *Í.F.* Ii, 32. See S. Nordal, *Íslenzk Menning* (Reykjavik, 1942), p. 54. Björn Thorsteinsson, *Ný Islandssaga* (Reykjavik, 1966), pp. 54–55.

17. *Lb.*, 325.

18. Alexander Bugge, *Vesterlandenes Indflydelse* (Christiania, 1905), pp. 366 sq.; Sveinsson, *Landnám i Skaftafellsþingi* (Reykjavik, 1948), pp. 23, 38–9, 41–2.

It is worth noticing that at this stage Morison abandons the evidence altogether and allows full rein to his powerful imaginative faculty. The whole of the lengthy excerpt set out below—with the single exception of the reference to Dicuil's account of the visit to

Iceland—is, in fact, pure invention. The allusion to 'the fleet of long Viking ships' is quite indefensible: there is no evidence whatever that any longship arrived in Iceland at the time of the Settlement. Again, the gratuitous introduction by this author of that absurd character, the 'stage Irishman', is a revealing example of Morison's chronic weakness for anachronism (*European Discovery of America*, pp. 26–7):

'In the late seventh and early eighth centuries, several hundred Irish ecclesiastics, mostly Culdees, were outraged by the Synod of Whitby's efforts to bring the Irish church into line with Rome. These clerics built curraghs and sailed to Iceland. There they were joined, from the close of the eighth century on, by Irish refugees from Viking raids on the Faeroes. Since Iceland had no indigenous population, the Irish made themselves at home, spreading out and forming monastics wherever it seemed possible to get a living. Dicuil, a learned Celt at the court of Charlemagne, told in a book that he wrote in 825, *De Mensura Orbis Terrae*, of hearing from a priest who had been to Iceland, a vivid description of the endless day in the Arctic summer, and used a homely example: one could even *pediculos de camisia extrahere*—pick lice out of one's shirt—as easily at midnight as at noon!

For a century and more, these Irish colonists in Iceland were safe from the tyranny of Rome, the rage of the Vikings, and the temptations of the flesh. By the second half of the ninth century, reinforced by subsequent waves of Celtic come-outers, they numbered at least several hundred, possibly more than a thousand. Alas, no part of the world could then ensure one a quiet life, any more than in our time. One fine day in the summer of 870, an alarming sight met Irish eyes on the Icelandic shore—a fleet of long Viking ships with attendant merchant vessels. One can imagine the consternation. What, in the name of Heaven, is this? asks one. Viking ships, m'lad; this is it! They are about to land! Mother of God, they are bringing cattle with them—and women—they mean to stay! Are those blue-eyed people with long yellow hair *women*, father? They are that, m'lad, and we shall have to leave this island to preserve our Christian purity!

The Norsemen kept coming, and the Irish *papae* or *papar*, as they called the monks, did not all leave at once. But within a few years they had all gone, and the Norsemen had Iceland to themselves.

Where did the Irish Icelanders go? Some, doubtless, went home to Mother Machree. Many must have perished at sea in their frail curraghs. But there is some reason to believe that a few sailed west, missed Greenland, and fetched up somewhere on the east coast of North America.'

19. *Lb.*, pp. 61–5. It is to be observed that towards the close of the eighth century Patrick, Bishop of the Sudreyjar, or Hebrides, showed himself remarkably familiar with Icelandic topography. Çf. *Lb.*, p. 53.

20. See 'Íslands- og Grænlandssiglingar og fundur Norður Ameríku á 15 öld', pp. 20–1, and *Enska öldin í sögu Íslendinga* (Reykjavik, 1970), pp. 262–3; Jón Jóhannesson, *Íslendinga saga* (Reykjavik, 1958), ii, 338–54.

21. *De mensura*, p. 74.

22. *Journal of Roman Studies*, vol. 24, p. 220.

23. Reeves, op. cit., pp. 40, 170–1.

24. Ibid., pp. 48–9.

25. Ibid., p. 168.

26. Niels Winther, *Færøernes Oldtidshistorie* (Copenhagen, 1875), p. 36. Winther's book, however, is based partly on Pastor Schroder's material, which is suspect.

27. *R.C.*, vol. 9, pp. 462–4; vol. 10, pp. 90–2. Cf. Reeves, op. cit., pp. 90–1. In early times mariners were closely observant of the habits of birds, which served them as one of the adventitious aids to navigation. There is another passage in the *Immram curaig Maíle Dúin* that seems to refer to some species of small auk. See *R.C.*, vol. 9, p. 475. They may also have been guided to some extent by the colour of the sea. See ibid., vol. 10, pp. 54–5.

28. See below, pp. 103–4.

29. In Adomnán's *Vita S. Columbae* there is an allusion to how 'the holy Bishop Colmán'

narrowly escaped drowning in a dangerous race known as Coire Brecáin, between Rathlin Island and the Irish mainland. See Reeves, op. cit., p. 29. This Coire Brecáin, under its modern appellation of the Slough-na-more race, is described in *The Irish Coast Pilot* as 'dangerous to small craft', etc. It was for this reason that in the early half of the nineteenth century the local coastguards used to receive extra pay.

30. *Lb.*, pp. 132–4, 395; *Í.F.* Ii, 245.

6. The genesis of Norse expansion

1. See P. H. Sawyer, *The Age of the Vikings* (London, 1962), pp. 66–82, 117–44; Gwyn Jones, *The Vikings* (London, 1969), pp. 182–203, 269–311.

2. There was an immense expansion of agricultural land in many parts of Scandinavia in the early medieval period. The multiplication of new farmsteads in Vest-Agder during the period 400–600 gives testimony of the steady growth of the population. See Magnus Olsen, *Farms and Fanes of Ancient Norway* (Christiania, 1928), pp. 110, 141; Haakon Shetelig, 'Norges Folk i Jernalderen', *Nordisk Kultur*, i, 46.

3. See Brøgger, *Ancient Emigrants*, pp. 18–19.

4. A. E. Christensen, *Boats of the North* (Oslo, 1968), p. 24.

5. It has sometimes been claimed for the Gokstad craft that every piece of work in the ship was admirably adapted to its own particular place and function and that all parts of the craft were in perfect balance and harmony with one another. The experience both of Captain Andersen in the *Viking*, a replica of the Gokstad vessel, which sailed from Norway to America in 1893, and of the crew of the *Hugin*, which made the passage from Denmark to the Kentish coast in 1949, would appear to support this contention to the full. See Magnus Andersen, *Vikingefærden* (Christiania 1895); Jorgen Røyel, *The 1949-Cruise of the Viking Ship 'Hugin'* (Copenhagen, 1949).

6. From time to time doubts have been expressed as to whether the side-rudder were really effective in a seaway. (See G. S. Laird Clowes, *Sailing Ships: Their history and development*, 1948, p. 43.) However, the practical experience derived from the voyages of the *Viking* and *Hugin* puts a rather different complexion on the matter. It is interesting to note that professional seamen who have had personal experience of the side-rudder hold a very different opinion about its efficiency. Captain Andersen remarked that it worked satisfactorily in every way and that one man could steer in any weather with simply a small line to help. Captain Sölver reported on the efficiency of the side-rudder fitted in the *Hugin* in very similar terms. The *Hugin*'s best point of sailing proved to be with the wind on her quarter. It was scarcely any effort to keep her on her course: the helmsman found he could steer this 30-ton craft with the utmost ease. Captain Sölver related that with the sail close-hauled the ship sailed within $5\frac{1}{2}$ to 6 points of the wind in a moderate breeze, making very little leeway; and that the rudder worked very satisfactorily on both tacks. See also Brøgger and Shetelig, op. cit., pp. 127–8.

7. A fitting for the *beitiáss* was also found in one of the Skuldelev ships ('Skuldelev I'): see Olaf Olsen and O. Crumlin Pedersen, 'The Skuldelev Ships', *Acta Archaeologica*, vol. 38 (1967), pp. 105, 107, 109. Cf. Harald Åkerlund, 'Áss och beiti-áss', *Unda Maris* (1956), pp. 30–1. A *beitiáss* was fitted in the *Hugin* in 1949, and gave very satisfactory results. Even when close-hauled, her speed was considerable and the leeway surprisingly small. Captain Sölver reported that the *Hugin* went about readily with a good working breeze in smooth water, but that in a seaway he had the oars on the lee side manned to bring her about, otherwise it would have been necessary to wear ship.

8. Brøgger and Shetelig, *The Viking Ships*, pp. 212–3.

9. Ibid., p. 185.

10. Brøgger, 'The Prehistoric Settlement of Northern Norway', *Bergens Museums Arbök* (1932), *Hist.-Antik. Rekke*, pp. 3–4. See also Shetelig, 'Et storre Norge' *Norsk Kulturhistorie*, Bd. i (1938), p. 237; Nils Vigeland, *Norge på Havet* (Oslo, 1953), p. 34.

11. It is not without significance that a pair of scales was discovered in the ship-burial at Kiloran Bay in Colonsay. See Shetelig, 'Ship-Burial at Kiloran Bay, Colonsay, Scotland', *Saga-Book*, vol. 5, p. 172.

12. *Lb.*, pp. 129, 162, 212, 260, 280, 328, 366; *Í.F.* xxvi, xxvii, *passim*; *Fb.* i, 130.

13. Shetelig, 'Et storre Norge', pp. 209–10, 215, 244, and also 'Vikinger i Vesterveg', *Nordisk Tidsskrift*, Årg. 9 (1933), pp. 422–3.

14. See Mohn, *Vindene i Nordsjøen ok Vikingetogen* (Christiania 1914), *passim*; Brøgger, *Ancient Emigrants*, pp. 24–5; Helland-Hansen and Nansen, *The Norwegian Sea* (1909), *passim*.

15. J. Jakobsen, *The Old Shetland Dialect* (Lerwick, 1897), p. 64, and also *Shetlandsøernes stednavne* (Christiania, 1901), pp. 220 sq.; Shetelig, *Préhistoire de la Norvège* (Oslo, 1926), Ch. 8, and also 'Vikinger i Vesterveg', pp. 422–3, 431; Brøgger, *Ancient Emigrants*, pp. 76, 118–21, 126–7.

16. 'Leabhar Oiris', ed. R. I. Best, *Eriu*, Vol. I.

17. Jackson, *Language and History in Early Britain*, p. 32.

7. The emigration to the Faeroe Islands

1. *Í.F.* xxvi, 118.

2. *De mensura orbis terrae*, ed. Tierney, pp. 74, 76.

3. *Lb.*, pp. 34–6. The late Professor Brøgger was of the opinion that Grím Kamban must have played much the same role in the emigration to the Faeroe Islands as Íngolf played in the settlement of Iceland and Eirík Raudi in that of Greenland.

4. This factor, *vestmannamotivet*, can be seen working in the *Landnámabók* and also in the *Færeyinga saga*. See Brøgger, *Norsk Geografisk Tidsskrift*, vol. 5, p. 239, and also *Løgtingssøga Føroya* (Oslo, 1935), pp. 98–9, 174.

5. Shetelig, *Préhistoire de la Norvège*, Ch. viii, and also 'Vikinger i Vesterveg', pp. 422–3, 431.

6. Brøgger, *Ancient Emigrants*, pp. 26–7.

7. Information given to the present writer by the late Professor Haakon Shetelig.

8. Sverrir Stove and Jakob Jakobsen, *Færøyane* (Oslo, 1944), p. 142.

9. See 'Rættarbót Hákunar hertuga Magnussonar', *Diplomatarium Færoense*, ed. J. Jakobsen (1907), pp. 3–22.

10. The word *knörr* is found in Old Danish as *knar*; in Faeroese, as *knörrur*; in Anglo-Saxon, as *cnear*; and in Old Irish, as *cnarr*. *Knörr* has often been rendered into English as 'cog', 'buss', and even 'galley'. All these interpretations are wrong and misleading, especially the last. The *knörr* was unquestionably a sailing ship, not a galley; for example, 'Byrr var a blásandi, ok gekk knörrinn brátt mikit . . . byrjaði honum vel til Ísland (*Fms.* vi, 249). For early references to the *knörr* see *Lb.*, pp. 70, 130, 136–8, 161, 163, 246, 281; *Í.F.* xxvi, 116; *Den norsk-islandske skjaldedigtning*, B, Bd.I, ed. F. Jónsson (Christiania, 1912), p. 415.

There were also vessels which did not fit readily into either of the broad categories of *langskip* and *hafskip*. For example, the imposing and superbly constructed Gokstad craft has been described sometimes as a fighting ship, and sometimes as a chieftain's vessel. In such cases anything like precise classification is out of the question. See Christensen, *Boats of the North*, p. 44.

11. See Vigeland, *Norge på Havet*, p. 30; G. J. Marcus, 'The Evolution of the *Knörr*', *M.M.*, vol. 41, pp. 119–20. The Åskekärr vessel, which was built in approximately 800, is an early example of the sea-going merchantman: see Humbla and Thomasson, 'Äskekärrsbåten', *Göteborgs och Bohusläns Fornminnesförenings Tidsskrift* (1934), pp. 1–34.

12. *Í.F.* xxvi, 118. Cf. *Ísl. sög.* ii, 98, 190, 277.

13. Cf. Winther, *Færøernes Oldtidshistorie*, p. 176; Hydrographic Dept., Admiralty, *Admiralty Weather Manual* (1941), p. 206. Captain Jacob Lund, for many years in the service of the Bergen Steamship Company, has informed the present writer that, when he was sailing between Hammerfest and Bear Island in 1904, it was observed that certain species of

seabirds were a useful guide to the mariner in thick weather, since when these birds 'got up from the sea we knew for certain that they would make for the bird rock on the western side of the mouth of the harbour'. Dr. Finn Devold, a well-known authority on the fisheries of his country, has similarly told the writer that Norwegian fishermen still use the flight-line of certain seabirds for locating Bear Island, the Faeroes, and Jan Mayen when approaching these islands in hazy weather.

14. Down to quite recent times many coasting and fishing skippers possessed this inborn knowledge and experience; and not a few deep-water sailors as well. Today, however, as a result of the latest scientific developments, this kind of skill is fast dying out. 'We go about the sea nowadays with our eyes shut,' an experienced Grimsby skipper once told the present writer. The whole matter of 'adventitious aids' (together with rule-of-thumb methods of nautical astronomy) were exhaustively examined, for life-saving purposes during the War, by Harold Gatty. See Harold Gatty, *The Raft Book* (New York, 1943), *passim*.

8. The settlement of Iceland

1. *Í.F.* ii, 5; *Lb.*, pp. 34–6; *Monumenta historica Norvegiae*, ed. Storm, pp. 8 *sq.*, 92. Cf. Sveinsson, *Landnám í Skaftafellsþingi*, pp. 41 *sq.*

2. *The Book of the Icelanders*, ed. H. Hermannsson (Ithaca, New York, 1930), p. 60.

3. Sveinsson, op. cit., pp. 121 *sq.*; B. Thorsteinsson, *Íslenzka þjóðveldið* (Reykjavik, 1953), pp. 78–85.

4. *Lb.*, *passim*. See Sveinsson, op. cit., p. 67 *et passim*.

5. *Lb.*, pp. 68, 70, 92, 125–6, 129, 212, 260, 272, 280, 366.

6. *Lb.*, pp. 36–7, 41, 50, 176, 181, 187, 197, 200–1, 216–17, 224–7, 244, 247, 270, 328. See also pp. 50, 180, 201, 387. Cf. Shetelig, 'Et storre Norge', p. 222; Sigurdur Nordal, *Íslenzk Menning* (Reykjavik, 1942), pp. 67–8, 71–3, 244; Sveinsson, op. cit., p. 47 *et passim*.

7. *Lb.*, pp. 50, 151–5, 176, 180, 186, 209, 217, 224–5, 379, 387.

8. *Lb.*, pp. 219–20, 285; *The Vatnsdalers Saga*, trans. Gwyn Jones (1944), p. 52; *Isl. sög. vii*, 40–1; *Í.F.* xxvi, 271.

9. *Lb.*, pp. 52, 55, 139, 187, 197–8, 200, 278, 319.

10. See Johan Schreiner, 'Slaget i Havsfjord', *Festskrift til Halvdan Koht* (Oslo, 1933), p. 103; J. Bing, 'Studier in Harald Harfagres og hans Sønners Historie', *Norsk Historisk Tidskrift*, Bd. 33, p. 12.

11. *Lb.*, pp. 176, 180, 186, 197–8, 200, 316–17, 343, 379.

12. *Hb.*, p. 462; *Fas.* i, 132.

13. *Fms.* iv. 280; xi. 182; *Færeyinga saga*, ed. C. C. Rafn (Copenhagen, 1832), p. 100. See also Brøgger and Shetelig, *The Viking Ships*, p. 234.

14. Mention is made in the *Ólafs saga helga* of the case of a *knörr* outsailing three pursuing longships: see *Fms.* vi. 249.

15. The Rebæk rudder, which was recovered by Dr. Pøul Nørlund and Captain Carl V. Sölver in 1943, is believed to have come out of a *hafskip*. It is fashioned from a very heavy piece of oak and measures over 13 feet in length. 'The Rebæk rudder apparently belonged to a merchantman of the Viking period or very little later, a *knörr* or a *byrðingr*, types which are known only from references in literature. This Rebæk rudder has a general resemblance to those of the Gokstad and Oseberg ships, but it differs in several respects, particularly in being much larger. . . . A side-rudder such as this is the most perfect thing of its kind and must have been the outcome of a long process of development.' See Carl V..Sölver, 'The Rebæk Rudder', *M.M.*, vol. 32, pp. 115–20.

16. *Fms.* ix. 45; *Ísl. sög.* v. 410.

17. Vessels of a much later date than this were fitted with oars for the same purpose. Some eighty years ago the last of the Greenland brigs, *Hvalposhen*, was thus equipped with four large oars.

18. Thus no attempt to use the oars was made on the occasion of the shipwreck off

Hornstrandir, on the north Iceland coast, in the *Gudmunðar saga*: see *The Life of Gudmund the Good*, ed. and trans. G. E. O. Turville-Petre and Olszewska (Oxford, 1942), p. 10.

19. See *Bps*. ii. 200.

20. The basin-shaped cog, which was later to supplant the Norse *hafskip*, though it had a superior carrying capacity, was by no means so fast a sailer.

21. *Lb*., pp. 68, 70, 130, 136, 163, 280; *Í.F*. xxvi, 116.

22. *Lb*., pp. 68, 70, 124–6, 138–40, 250–2.

23. *Lb*., pp. 68, 187, 197, 221, 229, 296, 327, 332, 346–7, 384; *Ísl. sög*. vii, 41.

24. Nordal, op. cit., pp. 53, 68, 72–3. The Icelandic sagas, which are so often full of anachronisms—for example, in the field of law—are usually sufficiently dependable in matters relating to shipbuilding, seamanship, and navigation. (Among the rare exceptions may be mentioned the well-known anachronism in the *Heimskringla* concerning the discovery of the Faeroe Islands; and there is another in connection with the Skania herring fishery.) The testimony of the sagas has been confirmed by that of far earlier sources and by the evidence of archaeology. The factor of *continuity* in these matters is, in fact, quite remarkable.

25. *Lb*., p. 68.

26. *Lb*., p. 70 *et passim*.

27. *Lb*., pp. 52, 187–8.

28. *Lb*., p. 43. See p. 137.

29. 'Kraku-Hreidar . . . kom i Skagafiord ok sigldi vppa Borgársand til brots' (*Lb*., pp. 232–3). See also pp. 188, 197.

30. 'þeim beit ei fyrer Reykianes' (*Lb*., p. 38). Cf. F. Jónsson, *De gamle Eddadigte* (Oslo, 1932), pp. 36, 187, 236, and also *Den norsk-islandske skjaldedigtning*, B, Bd.I, p. 500. It is to be noted that the blocks for securing the *beitiáss* were discovered many years ago in the Gokstad craft; though their true significance was not understood until comparatively recently.

31. *Lb*., pp. 186, 272, 280, 316–7, 328. Cf. *Hb*., p. 4; *Ísl. sög*. ii, 251; vi, 46–51; viii, 103–12. See also Vigeland, *Norge på Havet*, p. 39; Carl V. Sølver, *Vestervejen om Vikingernes Sejlads* (Copenhagen, 1954), pp. 22–39, 104–7.

32. See *Den norsk-islandske skjaldedigtning*, B, Bd.I, pp. 4, 668; ibid. B, Bd.II, p. 576. See also Hj. Falk, *Altnordisches Seewesen* (Heidelberg, 1912), pp. 19, 54; O. S. Reuter, *Germanische Himmelskunde* (Munich, 1934), pp. 598 sq.; Sølver, *Vestervejen*, pp. 131–2. But see also Schnall, *Navigation der Wikinger* (Hamburg, 1975), p. 90.

33. *Lb*., p. 52. See G. J. Marcus, 'Hafvilla: A Note on Norse navigation', *Speculum*, vol. 30, pp. 601–5.

9. The explorations of Eirík Raudi

1. *Í.F*. ii, 13–14; *Lb*., pp. 110, 131, 190–1, 195.

2. H. L. Sawatzky and W. H. Lehr, 'The Arctic Mirage and the Early North Atlantic', p. 1304.

3. Morison's belief that Gunnbjörn was a *friend* of Eirík's (*European Discovery*, p. 39) is without any foundation.

4. *Í.F*. ii, 13–14; *Lb*., pp. 133–5.

5. M. Thórdarson, *The Vinland Voyages* (1930), p. xv.

6. *Lb*., p. 132.

7. *Lb*., pp. 132–4, 395; *Í.F*. ii, 246.

8. L. Kristjánsson, 'Grænlenzki landnemaflotinn og breiðfirzki báturinn', *Árbók Hins íslenzka fornleifélags* (Reykjavik, 1964), pp. 64–8.

9. Jóhannesson, *Íslendinga saga*, i, 119; Bogi T. Melsted, *Ferðir, siglingar, og samgöngur milli Ísland og annara landa á dögum þjóðveldins* (Reykjavik, 1914), p. 647.

10. *The Book of the Icelanders*, ed. Hermannsson, p. 64.

10. The voyage of Bjarni Herjólfsson

1. Jón Jóhannesson, 'Aldur Grænlendinga sögu', trans. into English in Saga-Book, vol. 16, pp. 54–66.
2. G. M. Gathorne-Hardy, The Norse Discoverers of America (Oxford, 1921), p. 245.
3. See Saga-Book, p. 66.
4. Í.F. ii, 241–7.
5. Dœgr or dœgr-sigling, 'day's sailing', denotes an approximate measure of distance, not a period of time, as is explained in a later chapter.
6. Í.F. ii, 247; Gathorne-Hardy, op. cit., pp. 25–8.
7. Ibid., p. 250.
8. Í.F. ii, 246.

11. The settlement of Greenland

1. Middle Settlement is the name which has been given to this region in our own days; in all probability it formed part of the Eystribygd.
2. The King's Mirror, ed. and trans. L. M. Larson (New York, 1917), pp. 144, 149.
3. Ibid., p. 142.
4. Ibid., p. 143.
5. The Sagas of Kormák and The Sworn Brothers, ed. and trans. Lee M. Hollander (New York, 1949), pp. 145–8.
6. Adam of Bremen, Gesta Hammaburgensis Ecclesiae Pontificum ed. J. M. Lappenburg (Hanover, 1876), iii, 24, 73. In the early days of the colony, according to the same authority, Greenland was considered to be within the jurisdiction of the see of Skálaholt (Iceland).
7. Isl. Ann., p. 19.
8. Ísl. sög. i, 393–411.
9. Pøul Nørlund, Norse Ruins at Gardar (1930), pp. 47, 57, and also Viking Settlers in Greenland (1936), p. 34.

12. The Vinland voyages

1. Björn Thorsteinsson, 'Some Observations on the Discoveries and the Cultural History of the Norsemen', Saga-Book, vol. 16, p. 182.
2. See Í.F. iv, 248–69; Gathorne-Hardy, op. cit., pp. 49–72.
3. Adam received this information from Svein Estridsson, King of Denmark. 'Besides Iceland, there are many other islands in the great ocean, of which Greenland is not the smallest; it lies further away in the ocean. . . . It is said that Christianity has recently spread to them [the Greenlanders]. Moreover, he [the King of Denmark] said that an island has been found by many in this ocean, which has been called Vinland, because there vines grow wild and bear good grapes. Moreover, that there is self-grown grain in abundance, we learned, not from mythological tales, but from reliable account of the Danes' (Gathorne-Hardy, The Norse Discoverers of America, p. 98).
4. Í.E. ii, 13.
5. Lb., pp. 162, 241.
6. Symbolae ad geographiam medii ævi, ed. E. C. Werlauff (Copenhagen, 1821), p. 14.
7. Isl. Ann., pp. 19, 213.
8. Erik Wahlgren, The Kensington Stone (1958), p. 3.
9. See Hjalmar R. Holand, The Kensington Stone (Wisconsin, 1932) and also Westward from Vinland (New York, 1940). It may be mentioned in passing that Holand occupies something of the same position in relation to the Kensington Stone that W. L. S. Harrison does to the Titanic-Californian imbroglio. In both these cases it would seem to be a Cause rather than a controversy that is actually involved.

10. 'Never has a spurious document stood on such feeble ground and given such striking proofs of its falsity. If it has been allowed to live so long, that is because it has never before been subjected to a penetrating scholarly analysis' (Erik Moltke, 'The Ghost of the Kensington Stone', *Scandinavian Studies*, vol. 25, p. 14).

11. See the masterly survey of the whole matter of 'Norse antiquities' in America by Johannes Brønsted in *Aarboger for nordisk oldkyndighed og historie* (1950), pp. 1–151. This was, however, written long before Helge Ingstad's discoveries at L'Anse aux Meadows and the reports of Norse sites in the Ungava peninsula.

12. Helge Ingstad, 'The Norse Settlement at L'Anse aux Meadows, Newfoundland', *Acta Archaeologica*, vol. 41, pp. 109–54. See also his *Westward to Vinland* (1958), Pl. 32a, pp. 214–15, and *Land under the Pole Star* (London, 1966), p. 166. A caveat must nevertheless be entered against certain of the exaggerated claims advanced on the strength of these discoveries: claims which are demonstrably in excess of the evidence. Ingstad, indeed, appears to have ended by fully convincing himself—and Morison— that L'Anse aux Meadows was the real, veritable, and authentic Vinland where the Greenlanders had attempted to settle in the early eleventh century. 'So, now that the location is fixed', says Morison cheerfully, 'we may proceed with the story.' Thereafter he treats of what is no more than a bare possibility as if it were a proven fact. Thus: '*Mathew* [Cabot's ship] was now only five miles as the crow flies from L'Anse aux Meadows where Leif Ericsson had tried to establish a colony in 1001'. 'The site was used by two later and more populous expeditions from Greenland, which no doubt added to Leif's structures, making a small village' (*European Discovery*, pp. 38, 49, 172). A few aerial photographs of the alleged 'Wonder Strands' and other natural features mentioned in the sagas complete the story of the House that Leif (or Sam) Built.

13. Wilcomb E. Washburn, *Proceedings of the Vinland Map Conference* (Chicago, 1971), p. 126.

14. 'Some Observations on the Discoveries and Cultural History of the Norsemen', *Saga-Book*, Vol. 16, p. 182.

15. Skelton, R. A., Marston, T. E., and Painter, G. D., *The Vinland Map and the Tartar Relation* (New Haven, Conn., 1965), pp. 233, 257, 262.

16. *Isl. Ann.*, p. 213.

17. E. Reman, *The Norse Discoveries and Explorations in America* (Berkeley and L.A., 1949), p. 179.

18. Werlauff, op. cit., p. 14.

19. *Isl. Ann.*, p. 19.

20. Ibid., pp. 120, 213.

21. Gathorne-Hardy, op. cit., p. 297.

13. The Norse traffic to Iceland

1. *Isl. Ann.*, pp. 68, 360, 367, 414.

2. Ibid., p. 288, Cf. pp. 286, 358.

3. It seems possible that Professor Schreiner rather underestimates the quantity of heavy goods that could be transported in a *kaupskip*: see Johan Schreiner in *Norsk Historisk Tidsskrift*, Bd.36, p. 259.

4. See A. Bugge, *Den norske trælasthandels historie* (Skien, 1928), pp. 61 *sq.* Certain of the annals and sagas throw a good deal of light on the timber trade with Norway. See *Isl. Ann.*, pp. 342, 352; *Bps.* i, 26, 289; *Ísl. sög.* v, 12; vi, 343; viii, 155; ix, 149, 199.

5. *N.G.L.* iv, 292; *Grágás*, ii, ed. Finsen (Copenhagen, 1851), pp. 122–35.

6. *Fms.* vi, 266. Cf. *Bps.* ii, 216–7.

7. *Bps.*, passim; *Ísl. sög.*, passim.

8. *N.G.L.* ii, 440; *D.N.* xix, *passim*; *Bps.* i, 367, 395.

9. *D.N.* xix, no. 58.

10. See Ole Tonning, *Commerce and Trade on the North Atlantic, 850 A.D. to 1350 A.D.* (Minnesota, 1935), p. 6.

11. *D.N.* ii, no. 235; vii, nos. 103–4; A. Bugge, *Studier over de norske byers selvstyre og handel før Hanseaternes tid* (Christiania, 1899), pp. 114–15, 131 *sq.*, and also *Den norske sjøfarts historie*, pp. 234, 260; Knud Gjerset, *The History of Iceland* (1924), pp. 236, 244; E. M. Carus Wilson, 'The Iceland Trade', *Studies in English Trade in the Fifteenth Century*, ed. Power and Postan (1933), p. 157.

12. Especially the shipping news. See *Isl. Ann.*, pp. 267, 285, 418.

13. Ibid., pp. 23, 122–3, 130, 343, 384, 418. See also *D.N.* i, no. 25; *Bps.* ii, 291.

14. *Isl. Ann.*, pp. 288, 356.

15. Ibid., pp. 339, 358, 360; *Bps.* i, 14; *Ísl. sög.* i, 422; v, 259–60.

16. *Ísl. sög.*, i, 268, 285, 360, 388, 412.

17. Ibid., p. 388; *Bps.* i, 14–15; *Ísl. sög.* ix, 34; *Sturls.* i.

18. See *Lb.*, pp. 35, 39; *Isl. Ann.*, p. 292; *Bps.* iii, 252, 346; *Sturls.* ii, 85; *Ísl. sög.*, xi, 192. Cf. Vigeland, *Norge på Havet*, pp. 36, 39.

19. *Ísl. sög.* ii, 69, 75; vi, 401. See Hydrographic Dept., Admiralty, *Arctic Pilot*, ii (1949), pp. 62–6.

20. *Lb.*, pp. 220, 285; *The Vatnsdalers' Saga*, ed. and trans. Gwyn Jones, p. 52. The northabout route would also be followed under certain weather conditions; see *Bps.* ii, 290.

21. *The Life of St. Lawrence*, trans. O. Elton (1890), pp. 126 *sq.* Cf. *Bps.* i, 395; ii, 221.

22. The record number of sailings from Norway to Iceland seems to have been thirty-five, which occurred in the year 1118: see *Bps.* i, 14.

23. *Isl. Ann.*, pp. 183, 189, 206; *Fms.* vi, 239; *Sturls.* ii, 363.

24. *Isl. Ann.*, pp. 200, 227, 274, 366; *The Life of Gudmund the Good*, ed. and trans. Turville-Petre and Olszewska, pp. 17, 55; *Sturls.* ii, 416. Cf. ibid. 493.

25. See *Isl. Ann., Bps., Sturls., Ísl. sög., passim*. These vessels were often named after saints, like *Maríusúðin, Óláfssúðin, þorlákssúðin*, and *Benediktsbátin*; and sometimes after some cleric.

26. *Sturls.* i, 224.

27. *Isl. Ann.*, pp. 154, 343–4, 348.

28. Ibid., pp. 222, 349.

29. Ibid., pp. 361, 366.

30. Ibid., pp. 280, 363.

31. This estimate is deduced from both documentary and archaeological evidence; notably the Skuldelev ships.

32. *Fms.* vi, 31, 36; *Ísl. sög.* iv, 467; v, 7; x, 8, 28, 164–5.

33. See *N.G.L.* ii, 277; *Isl. Ann.*, p. 267; *Bps.* i, 373; ii, 290; *Sturls.* ii, 363; iii, 106; *Ísl. sög.* i, 421; ii, 94; v, 248; vii, 392; x, 267; xi, 202, 205, 411; *Fms.* vi, 379; *Fas.* iii, 173, 371.

34. *Norske gaardnavne*, ed. Rygh, *passim*. Cf. Knarrarstaðir, now Knarstoun in the parish of Orfor, Orkney Islands: see *Fb.* ii, 439.

35. *Fms.* v, 74.

36. *Fms.* ix, 38; x, 400; *Sturls.* iii, 8.

37. For the use of the *knörr* on the ocean trade-route, see *D.N.* vii, nos. 103–4; *Isl. Ann.*, pp. 69, 112, 212, 228, 293–4, 361; *Bps.* i, 339; *Hb.*, pp. 131, 500; *Sturls.* ii, 71; *Ísl. sög.* iii, 178; iv, 60; v, 239; vi, 125; vii, 375, 392; viii, 128, 401; ix, 149; x, 123; *Fms.* i, 246; ii, 107; vi, 198, 239, 305, 377; ix, 292.

38. According to the *Byskupa sögur*, the *búza* was apparently used in the Norway–Iceland traffic; from Germany to Bergen, Bishop Gudmund travelled in a cog: see *Bps.* iii, 382. Also there are occasional references to the use of the *knörr* on the Iceland trade-route as late as the fifteenth century: see *Isl. Ann.*, pp. 293–4.

39. *Fb.* ii, 256; *Fms.* iv, 297.

40. *Isl. Ann., passim*; Bps. iii, 241, 246, 346; *Sturls., passim*. Cf. *D.I.* ii, nos., 464, 485. A *búza*, called the *Vallabúzuna*, is first heard of in 1184, on the Iceland run.

41. *D.N.* xix, no. 436; *Sturls.* ii, 71; *Í.F.* xxvii, 195, 236–7; *Fms.* ix, 20, 45; *Ísl. sög.* iv, 414; v, 15. See also Falk, op. cit., pp. 108 *sq.*; Brøgger and Shetelig, op. cit., pp. 234, 236.

42. Falk and Shetelig, *Scandinavian Archaeology* (Oxford, 1937), p. 364.

43. Sölver, *Om Ankre* (Copenhagen, 1945), p. 65.

44. See *Ísl. sög.* ii, 75.

45. An iron anchor nearly three times the size of the Oseberg anchor, and of approximately the same date, was discovered at Dønna in the Nordland. It is preserved in the Sjøfartsmuseum at Bergen.

46. H. Falk and H. Shetelig, *Scandinavian Archaeology* (Oxford, 1930), p. 369; *Om Ankre*, pp. 43, 46.

47. Sölver, *Acta Archaeologica*, vol. xvii, pp. 122, 126. There are a number of passages in the sagas which give testimony of the toughness of the Norse ground-tackle: see *Bps.* ii, 203 *et passim*; *Icelandic Sagas*, ed. and trans. R. G. Dasent (1894), iv, 356. See also *N.G.L.* ii, 280–2.

48. *Lb.*, pp. 219–20; *Ísl. sög.* ii, 37–8; vii, 40–1.

49. See Olaf Olsen and Ole Crumlin Pedersen, 'The Skuldelev Ships', *Acta Archaeologica*, vol. 29, pp. 161–75, and also vol. 38, pp. 73–174. The Skuldalev wrecks, according to Morison, 'have immensely increased our knowledge of the merchant ship of the period' (*European Discovery*, p. 63). This is certainly an exaggeration. The Skuldelev evidence for the most part only confirms what was already known from literary sources. See also D. Ellmer, *Frühmittelalterliche Handelsschiffahrt* (Neumünster, 1972), pp. 39, 46,

50. 'The Skuldelev Ships', pp. 108–9.

51. *Ísl. sög.* xi, 210.

52. Elton, op. cit., p. 52.

53. *Bps.* i, 59. See also ii, 186–7, 466, 479; *Ísl. sög.* iv, 341; v, 99; vii, 294, 298; x, 27, 267, 314.

54. It is related that in the reign of Óláf the Saint a strong east wind carried Thórarinn Nefjólfsson and his ship across the Iceland Sea from Norway in the record time of four days and four nights (*Íst. sög.*, ix, 161). Cf. vii, 294.

55. Elton, op. cit., p. 44.

56. See *Sturls.* ii, 85; *Fms.* vi, 377; *Ísl. sög.* ii, 82; iii, 135; iv, 330; v, 7, 14, 120, 283; ix, 133, 135, 158; x, 121.

57. *Isl. Ann.*, pp. 226, 280, 343, 346, 348, 365, 367.

58. Ibid., pp. 278, 478; *Ísl. sög.* xi, 434; *Sturls.* iii, 466; *Bps.* i, 14.

59. See *Isl. Ann.*, p. 290.

60. For example, it is apparent that not all the losses on the Greenland coast were recorded, as may be seen by comparing a certain passage in *Konungs Skuggsjá* with the list of contemporary wrecks in the Icelandic annals. Cf. *Ksk.*, p. 28; *Isl. Ann., passim*. It was estimated by Dr. Hasund that, on average, a shipwreck occurred every two years during the fourteenth century according to the Icelandic annals; but it is doubtful whether these figures can be accepted for the reasons already given. See S. Hasund, *Det. norske folks liv og historie gjennem tidene*, iii (Oslo, 1934), pp. 197–8.

61. *Isl. Ann.*, pp. 112, 180, 362, 367; *Bps.* i, 14–15.

62. See *Isl. Ann.*, p. 401: 'A ship ran on Thioraasand with the loss of eighteen men. Snorri's ship was wrecked and he and five others were drowned. The *Lysubúza* was lost off Eyrar. The *Apostle* was damaged and later drove on Seal Island. The *Anguelldznesbúza* went to pieces in Grindavík. Two men were lost. The *Katrinarsúðina* was wrecked in Nyiahraun.'

63. See ibid., p. 208: 'Eleven merchantmen wintered in Iceland and a twelfth was wrecked off Eyri in the autum'; p. 212: 'There came to Iceland eleven ships one of which was lost off Sléttu and there were six in the Austfjörd'; p. 280: 'There came six ships to Iceland and nine had intended to do so'; p. 403: 'There came thirteen ships from Norway to Iceland'.

64. Turville-Petre and Olszewska, op. cit., p. 11. For navigational conditions on the Strandir coast, see *Arctic Pilot*, ii, 297.

65. Elton, op. cit., pp. 84–5.

66. *Ísl. sög.* xi, 434.

67. *Bps.* i, 59.

68. *Isl. Ann.*, p. 221.

69. *Ísl. sög.* x, 178.

70. *Isl. Ann.*, p. 348.

71. Ibid., p. 386. Cf. Professor P. G. Foote, *On the Saga of the Faeroe Islands* (London, 1964), pp. 10–11.

72. *Isl. Ann.*, p. 415.

73. *Bps.* i, 346.

74. *Isl. Ann.*, p. 267.

75. Ibid., p. 389.

76. Ibid., p. 220.

77. Ibid., p. 69.

78. Ibid., p. 340.

79. *Fms.* ix, 28.

80. *Ísl. sög.* v, 14.

81. *Isl. Ann.*, p. 400.

82. Ibid., pp. 187, 193, 280, 345, 359, 390.

83. Ibid., pp. 364–5, 413–14.

84. In the fourteenth century alone there are many such examples. See ibid., pp. 152, 267, 285, 350, 365, 402, 421, 482; *Bps.* iii, 346.

85. A number of such disasters occurred during the period 1306 to 1385: see pp. 148, 201, 215, 286, 358, 414.

86. See *Ísl. sög.* ix, 163. Cf. Thorsteinsson, *Íslenzka þjóðveldið*, pp. 187–90.

87. Ibid., pp. 306–12.

88. *D.I.* i, no. 152; *Den norske sjøfarts historie*, pp. 187, 200, 233 *sq.*, 260; Gjerset, op. cit., p. 206; Thorsteinsson, op. cit., pp. 313–16. A similar arrangement for the provision of shipping was made with the Faeroe Islands: see *Isl. Ann.*, p. 360.

89. Gjerset, op. cit., pp. 235 *sq.*

90. *Den norske sjøfarts historie*, pp. 234–5; Gjerset, *op. cit.*, pp. 240–1; L. Musset, *Les Peuples scandinaves au moyen âge* (Paris, 1951), p. 74.

91. *Isl. Ann.*, pp. 208, 211–12, 401, 403.

92. The Archbishop of Nidarós was privileged to export thirty lasts of flour each year to Iceland: see *N.G.L.* i, 459, 465, 472.

93. Schreiner has shown that conditions in Norway in this era were unfavourable to the rise of a considerable mercantile class and that the share of the merchants in the seaborne trade of the realm was not very large. Bishops, monks, and nobles trafficked largely on their own account. See Johan Schreiner, *Norsk Historisk Tidsskrift*, Bd.36, p. 263.

94. Gjerset, op. cit., p. 244.

95. In the second decade of the fourteenth century several of the Norwegian vessels which formerly sailed to England were now voyaging to Iceland. In the 1340s there were others, among them the *Grandabúza* (of Nidarós), the *Shankinn* (of Oslo), and the *Avalsnesbúza*. Towards the end of the century there 'was also a *búza* belonging to the monks of Halsnøy in South Hordaland. See *Isl. Ann., passim*.

96. It is recorded in the Icelandic annals that, in the summer of 1294, the volcanic ashes, *suartir flakar*, proceeding from one such eruption were observed by certain merchants as far off as the vicinity of the Faeroes (ibid., p. 485).

97. See Hasund, op. cit., pp. 136–7.

98. *Isl. Ann.*, pp. 354–6, 403.

99. Gjerset, op. cit., p. 265, n. 1. See also *D.I.* iii, nos. 311, 315.

100. *Isl. Ann.*, pp. 365, 367, 412, 415–16, 424.

101. Certain of Bugge's conclusions on these points need to be drastically revised in the light of subsequent research. (This also applies to the contributions to Icelandic history of Knut Gjerset and Professor E. M. Carus Wilson.) See Maria Wetki in *Hansische Geschichtsblätter*, Bd.70, p. 43; Johan Schreiner in *Norsk Historisk Tidsskrift*, Bd.36, pp. 259–65.

14. The Greenland trade-route

1. It is worth noticing that a number of mill-stones have been discovered in the East Settlement, especially around Gardar. Though it is possible that the Greenlanders received cargoes of grain from Europe, from which they ground their flour, it is, on the whole, probable that it was a meal which was actually imported to Greenland; in view of the fact that it was flour, rather than grain, which was generally imported to Iceland.

2. It is probable that some of this timber came from North America, which was by far their nearest source of supply. Karlsefni and others, it will be remembered, had brought back timber to the Eiríksfjörd from the American coast; and a voyage was made to 'Markland' as late as 1347. See *Fb.* i, 540, 548; *Isl. Ann.*, p. 213; *G.H.M.* iii, 243.

3. *Ksk.*, p. 29.

4. Larson, *The King's Mirror*, p. 142. See Finn Gad, *Grønlands historie*, i (Copenhagen, 1967), pp. 152–3, 155–6.

5. *G.H.M.* iii, 242.

6. *Ísl. sög.* i, 395–6.

7. *G.H.M.* iii, 243.

8. *Hauksbók*, ed. F. Jónsson (Copenhagen, 1892), p. 501. See Magnus Olsen, 'Kingingtorssuaqstenen og sproget i de grønlandske runeinnskrifter', *Norsk tidsskrift for sprogvidenskap*, vol. 5 (1932), pp. 189 *sq.*, 229; Ingstad, *Land under the Pole Star*, pp. 88–9.

9. Storm, *Monumenta historica Norvegiae* (Christiania, 1880), pp. 193, 258, 330. It will be observed that while the carrying capacity of the Greenland *knörr* was comparatively small, the cargoes transported were often of considerable value: see *Isl. Ann.*, p. 212; *Ksk.*, pp. 29–30; *Ísl. sög.*, i, 393; iv, 434, 446; *Hauksbók*, p. 500.

10. *D.I.* i, no. 71.

11. *Pavelige Nuntiers Regnskabs- og Dagböger förte under Tiende Opkrævningen i Norden, 1282–1334*, ed. P. A. Munch (Christiania, 1864), pp. 25–8. It is to be noted that the tithes from all the other Scandinavian sees were paid in money: from Greenland alone were they paid in kind. See ibid., *passim*.

12. Larson, op. cit., p. 144.

13. See Müller-Röder, *Die Beizjagd und der Falkensport in alter und neurer Zeit* (Berlin, 1906), pp. 2, 10–11, 15.

14. *Ksk.*, pp. 30, 138; *Bps.* i, 4; *Fb.* i, 445 *sq.*; *Ísl. sög.* i, 410; *Fms.* vi, 298 *sq.* According to the *Finnboga saga*, Earl Hákon possessed 'a white bear' that could understand human speech.

15. *Ísl. sög.* i, 368, 389, 393; iv, 426, 430, 434–5; v, 280, 326, 389. Most of the overseas traffic of Greenland was with Norway; but a number of voyages, especially in the earlier centuries, were made to Iceland. There was at this time a fairly close connection between the two lands. The fame of St. Thorlák, it is stated, extended to Greenland as to other parts of the Norse world. From time to time a bishop of Gardar would visit his brethren of Skálaholt and Hólar. But these Iceland voyages, by all accounts, were few and far between.

16. Ibid. i, 395–6.

17. 'Fengu þeir enn veðr andstæð, ok velkti þá norðí Grænlandshaf. Fóru þeir svo nær Íslandi, at þeir sáu jöklana. Helgi gekk aldri frá stýri. Drifu þeir enn lengi (ibid. i, 428). See also iv, 439; v, 289. Long passages of this kind were by no means infrequent on the Greenland route.

18. *Í.F.* xxvii, 127.

19. *Isl. Ann.*, pp. 123–4.

20. *Ísl. sög.* i, 395 *sq.*

21. *Isl. Ann.*, pp. 136, 357.

22. Ibid., p. 228 *et passim.*

23. Larson, op. cit., pp. 101, 137–8; *Lb.*, p. 124. See I. J. Steenstrup, 'Hvad er Kongespeilets Havgjerdinger', *Aarbøger for nordisk Oldkyndighed og Historie* (1871).

24. *Isl. Ann.*, *passim*; Ksk., pp. 28–9.

25. *Ksk.*, pp. 27–9; *D.N.* i, no. 66; v, no. 152; *Ísl. sög.* ix, 157.

26. See Professor H. A. L. Fisher, *History of Europe* (1957 edn.), p. 180; Professor E. M. Carus Wilson, 'The Iceland Trade', p. 158; Professor J. W. Thompson, *Economic and Social History of the Middle Ages* (1928), p. 282; Mr F. W. Brooks, *English Naval Forces, 1199–1272*, p. 4; Lieutenant-Commander P. K. Kemp, *The Oxford Companion to Ships and the Sea* (1975), p. 497.

27. *Ísl. sög.* i, 381, 393, 397; *Sturls.* ii, 71; *Hauksbók*, p. 500. See also Gad, op. cit., pp. 151–2. Some ship's planks of pine found near Sandes in Greenland by Aage Rousell were very similar to the planks of the Skuldelev *knörr*. See *M.O.G.*, vol. 88, no. 2.

28. *Isl. Ann.*, p. 196.

29. In the Icelandic sagas the distinction between the *hafskip* and *langskip* is clearly expressed: see *Fms.* vi, 360; *Ísl. sög.* vii, 146; xi, 202. Cf. *Ísl. sög.* ii, 37; viii, 126. Cf. also *Jónsbók*, ed. Ól. Halldórsson (Copenhagen, 1904), pp. 256–8. *Ksk.*, pp. 28, 42. Longships frequently made the shorter crossing to the Scottish isles: see *Fms.*, *passim.*

30. *Í.F.* xxvi, 336. In the case of the Gokstad vessel the freeboard amidships is less than $3\frac{1}{2}$ feet: see Brøgger and Shetelig, *The Viking Ships*, p. 112.

31. *Fms.* iv, 284; *Í.F.* xxvii, 216; *Fær. s.*, p. 100. See also *Ksk.*, pp. 28, 42. The idea that the Viking longship, or *langskip*, 'could ride out the fiercest storms of the Atlantic Ocean' is an illusion which some time ago received wide currency in Sir Winston Churchill's *History of the English-Speaking Peoples* (1956), i, 72.

32. *Hb.*, p. 4.

33. There are a number of passages in the Icelandic sagas, notably in the *Grettis saga*, which strongly suggest latitude sailing.

34. Dating from the mid-fourteenth century is a set of sailing directions for the Greenland coast, which was long afterwards translated into English for the use of Henry Hudson. It is probable that parts of these sailing directions were in use at a much earlier date. This detailed description of the long, fjord-indented southern coast of Greenland, skirted by islands, skerries, and outlying dangers, served as a guide to the mariner in those distant waters, at a time when charts were, as yet unknown in the North. See *Det gamle Grønlands beskrivelse af Ívar Bárðarson*, ed. F. Jónsson (Copenhagen, 1930), pp. 18 *sq.* Ivar Bárdarson's original book, written in Old Norse, is no longer extant; but there are copies in Danish and other languages.

35. *Hb.*, p. 4; *Det gamle Grønlands beskrivelse af Ívar Barðarson*, p. 19; Poul Nørlund, 'Buried Norsemen at Herjolfsnes', *M.O.G.*, vol. 67, p. 236.

36. *Scriptores Rerum Danicarum medii aevi*, ed. J. Langebek (Copenhagen, 1778), v, 352.

37. *D.N.* vii, nos. 103–4; Bugge, *Studier over de norske byers selvstyre og handel før Hanseaternes tid*, pp. 131–4.

38. *D.N.* ii, no. 235.

39. The reference is to 'Ólaf i Lexo oc Æindrida Arnason oc adra þeirra kumpana': see ibid.

40. *Fms.* x, 111; Vigeland, *Norge på Havet*, p. 51; Gad, op. cit., p. 156.

41. *Isl. Ann.*, p. 120.

42. *D.N.* ii, no. 66; v, no. 152.

43. *Lb.*, p. 32.

44. *Flóamanna saga*, ed. F. Jónsson (Copenhagen, 1932), p. 34.

45. *Isl. Ann.*, pp. 120, 181. Cf. *Sturls.*, p. 106.

46. Larson, op. cit., p. 138.
47. Ibid., pp. 138–9.
48. *Det gamle Grønlands beskrivelse*, p. 19.
49. *D.N.* v, no. 152.
50. 'Herra Alfr byskup kom til Grænlandz hafde þar þı uerid byskups laust vm xix ár' (*Isl. Ann.*, p. 228).
51. Ibid., p. 414.
52. *D.N.* vii, no. 103.
53. *Isl. Ann.*, passim.
54. Ibid., p. 212.
55. *G.H.M.* iii, 121–2.
56. See *Isl. Ann.*, p. 228: 'The Greenland *knörr* went down off Norway but all the men came safe ashore.'
57. *D.N.* xvii (B), pp. 280–3. See G. J. Marcus,'Gardar: the Diocese at the World's End', *I.E.R.*, vol. 79, pp. 99, 105–6.
58. *G.H.M.* iii, 125, 139–40.
59. See *Det gamle Grønlands beskrivelse*, pp. 30–1; *M.O.G.*, vol. 67, p. 248; vol. 76, p. 129.
60. The suggestion has been made that the Greenlanders may have overworked the walrus fishery: see Hasund, *Det norske folks liv og historie*, iii, 199.
61. *Isl. Ann.*, p. 361.
62. Ibid., p. 228.
63. Ibid., p. 414.

15. Seamanship and navigation

1. *N.G.L.* ii, 273. Cf. 273–5; *Ísl. sög.* vii, 223.
2. *Early Norwegian Laws*, ed. and trans. L. M. Larson (New York, 1935), p. 125. Cf. Sverre Steen, 'Veier og Reiseliv', *Norsk Kulturhistorie*, Bd.1 (1938), p. 389. See also *Bps.* i, 59; *Ísl. sög.* xi, 434. Cf. vii, 400; ix, 289.
3. *Jónsbók*, ed. Halldórsson, pp. 240, 259. See also *Isl. Ann.*, p. 367; *Sturls.* i, 16, 405; ii, 493; *Hb.*, p. 131; *Ísl. sög.*, ii, 251; iii, 125; iv, 79; v, 255; vii, 298; x, 82, 121; xi, 16.
4. *The King's Mirror*, ed. and trans. Larson, p. 83.
5. Some of the commoner accidents at sea are recorded in the early Norwegian laws. See *N.G.L.* ii, 282–4. Cf. *Fms.* vi, 381; viii, 209; ix, 314; *Sturls.*, p. 85; *Ísl. sög.* iii, 440–1; vii, 146; ix, 290; *Fb.* iii, 146; *Fas.* iii, 95, 354. It is interesting to notice that most of the Skuldelev wrecks bear signs of extensive repairs.
6. *The King's Mirror*, pp. 157, 161.
7. *Fb.* ii, 205.
8. *N.G.L.* ii, 274–5; *Ksk.*, p. 6. Cf. *Í.F.* xxvii, 236–7, 290.
9. Morison writes: 'Nothing is said in the literature of the Roskilde ships about the ballast' (*European Discovery*, p. 36). He would find no lack of references to *grjót* (ballast) in Norse shipping if he only looked in the right places. See *Lb.*, p. 70; *Bps.* i, 345; *Ísl. sög.* ii, 62; v, 236; *Fms.* ix, 44–5; *Fas.* i, 321.
10. In the *Gísla saga* there is a revealing passage which relates how a ship lay under some islands while waiting for a fair wind. 'Halda þeir nu sidan skipinu til hafs ok leggja þar til hafnar einnar undir eyjar nokkurar ok aetla þadan at bida byrjar': see *Ísl. sög.* v, 120. Cf. *Lb.*, p. 68; *Bps.* ii, 290; *Sturls* ii, 363; iii, 106; *Ísl. sög.* ii, 63; iii, 498; iv, 467; v, 25, 237; vi, 437; xii, 110; *Í.F.* xxvii, 236.
11. *Isl. Ann.*, p. 267; *Bps.* iii, 240; *Ísl. sög.* iv, 289; v, 408–10; vi, 437; *Fms.* vi, 382–3; *Fas.* iii, 381. See also the admirable paper by D. L. Binns, 'Ohthere's Northern Voyage', *English and Germanic Studies*, 7 (Birmingham, 1957) pp. 43–5; Schnall, *Navigation der Wikinger* (Hamburg, 1975) p. 180.

12. *The King's Mirror*, pp. 157, 161.

13. Ibid., p. 79.

14. *N.G.L.* ii, 275–6; *Jónobók*, p. 256. See also *Bps.* ii, 200; *Isl. šg.* v, 288. Cf. x, 324; *Gs.*, p. 246.

15. *Bps.* ii, 198, 292.

16. *Fms.* vi, 360.

17. *Ísl. sög.* iii, 103. There is a brief description of the work of watering ship in the *Gudmunðar saga*: see *Bps.* ii, 288. Occasionally crews are recorded to have been in dire straits through lack of water. See *Lb.*, p. 210; *Isl. Ann.*, pp. 294, 358.

18. Falk and Shetelig, *Scandinavian Archaeology*, p. 349; *Early Norwegian Laws*, p. 191; *N.G.L.* ii, 274–5. Haakon Shetelig, 'Et storre Norge. Hav og Sjøfart', *Norsk Kulturhistorie*, i (1932), p. 244.

19. *N.G.L.* ii, 276–7.

20. *Bps.* iii, 384. Cf. *Ísl. sög.* vii, 404; *Fas.* iii, 371. Cf. also *Fms.* x, 53; *Fb.* iii, 384; *Fas.* ii, 496; iii, 167, 275, 299. For the *vindáss* in the fifteenth-century Kalmar ship see Sölver, *Om Ankre*, pp. 65, 116.

21. *Fms.* iii, 26; x, 71; *Orks.*, p. 243; *Fb.* i, 540; ii, 482; *Ísl. sög.* i, 429; ii, 83, 441; iii, 467–8; vi, 17; vii, 40; ix, 85; *Fas.* iii, 154. It is to be observed that apparently not all vessels were able to tack. Cf. *Fb.* iii, 146. See also Brøgger and Shetelig, *The Viking Ships*, pp. 149–50; Vigeland, *Norge på Havet*, p. 115.

22. *The Life of Gudmund the Good*, ed. and trans. Turville-Petre and Olszewska, p. 56. Cf. *Bps.* ii, 171, 198, 290; iii, 244, 252; *Fas.* iii, 95. Cf. also *Alexanders saga*, ed. C. R. Unger (Christiania, 1848), p. 67. Similarly, on approaching the coast in the hours of darkness, a crew might lower the sail and allow the vessel to drift. See *Fms.* ii, 64. Cf. Schnall, *Navigation der Wikinger*, p. 107.

23. Turville-Petre and Olszewska, op. cit., p. 57.

24. See *Orks.*, p. 244: 'svidris vid raa midia'.

25. Turville-Petre and Olszewska, op. cit., p. 58.

26. *Orkneyinga saga*, ed. and trans. A. B. Taylor, p. 282. Cf. *Í.F.* iv, 246; *Fms.* v, 337; vii, 67; ix, 20–1; *Fb.* ii, 308; *Fas.* i, 35; iii, 95.

27. 'En í því kemr áfall sva mikit, at yfir gekk þegar skipit ok ofan drap flaugina ok af vígin bæði' (*Bps.* ii, 201). Cf. *Den norsk-islandske Skjaldedigtning*, ed. Jónsson, B, Bd.1 (1912), p. 413; *Fms.* ix, 44; *Fas.* ii, 77.

28. *Fas.* i, 298: 'Nú fœrðu þeir þvergirðinga à skipin'. See also Falk and Shetelig, op. cit., p. 350.

29. *N.G.L.* ii, 278. Cf. *Ísl. sög.* vii, 400.

30. *Fms.* x, 136: 'æsti storminn svá, at sumir hjoggu tréin'. Cf. *Fas.* i, 515.

31. *Ísl sög.* vi, 50–1.

32. *Fb.* ii, 204–5.

33. *N.G.L.* i, 335. Cf. *Fms.* viii, 139; *Ísl. sög.* iii, 408, 469; *Fas.* ii, 243; iii, 354.

34. *Ísl. sög.* vi, 50; *Fb.* ii, 204.

35. There is an excellent example of this in the *Illuga saga Tagldarbanar*. '.... ok þá harðast lík, berr þá under land, þar er fyrir vóru boðar ok blindsker. Veðr var þá dimmt ok ill syn. Lýkr svo at skipt berr á land ok brotnar til ófjœris. Menn komast allir lífs af, ok mastöllum fjárlut, fá þeir bjargat' (*Ísl. sög.* iii, 470). Cf. *Isl. Ann.*, pp. 152, 267, 285, 350, 365, 401–2, 421, 482; *Bps.*, iii, 346.

36. Taylor, op. cit., p. 277.

37. *The Story of Egil Skallagrimsson*, trans. W. C. Green (1893), p. 125.

38. *Ísl. sög.* ii, 69.

39. *Orks.*, p. 27.

40. Pentland Firth lay on one of the major sea-routes of the north; namely, that between Norway and the Hebrides and Ireland. To the small sailing craft of those days it was a perennial menace. 'The extreme violence of the race, especially with westerly or north-

westerly gales, can hardly be exaggerated. . . . Only personal experience can make anyone realise what a seething cauldron of tide-rips, whirlpools, and short choppy sea the firth can become under these conditions. Most probably a small steamer of slow speed, or a sailing vessel, would never have been heard of again' (Hydrographic Dept., Admiralty, *North Sea Pilot*, Pt. II, p. 66). See *Fb.* ii, 258: 'lægdu þeir segl ok köstudu ackerum ok bidu þeir straumfallz þuiat raust mikil var fyrir þeim'. Cf. i, 227; ii, 457, 497; iii, 175–6, 228.

41. Green, op. cit., p. 59. In the Icelandic sagas there are many such passages illustrating this cautious approach to the land. See *Ísl. sög.* ii, 55, 69; iii, 440–1, 452–3, 468; *Fas.* iii, 354; *Bps.* iii, 245, 252.

42. *Laxdœla saga*, trans. T. Veblen (New York, 1925), p. 12.

43. Turville-Petre and Olszewska, op. cit., p. 58. For navigational conditions in the vicinity of Sanda Island see Hydrographic Dept., Admiralty, *West Coast of Scotland Pilot*, pp. 60 sq.

44. *Ísl. sög.* viii, 104.

45. Veblen, op. cit., pp. 46–7.

46. The reason why the academic tribe have found this problem insoluble is, in all probability, because they never sought the views of professional seamen. Experienced master mariners might well have shed light on the matter; and ultimately (as is explained in the Preface) one of them did.

47. In Sigfús Blöndal's *Íslensk-Dönsk Orðabók* (1924) *hafvilla*, a feminine noun, is translated into Danish as 'Forvildelse til Søs, det at komme ud af Kursen', and in Fritzner's *Ordbog over det gamle norske Sprog* (1886) as 'Stilling hvori en Søfarende er, naar han ikke har Rede paa, hvor han er paa Havet'. In Zoëga's *Concise Dictionary of Old Icelandic* (1910) it is rendered into English as the 'loss of one's course at sea', and *hafvillr*, the adjective, as 'having lost one's course at sea'. The same renderings are given in Cleasby and Vigfusson's *Icelandic– English Dictionary* (1874).

48. *Í.F.* iv, 246; *Bps.* ii, 172; *Ísl. sög.* iii, 450; ix, 294; xi, 186.

49. *Fms.* iii, 181. Cf. *Fas.* iii, 402.

50. *Lb.*, p. 135.

51. *Ísl. sög.* ix, 294–5. Cf. *Fb.* i, 332; *Í.F.* iv, 246.

52. *Ísl. sög.* iv, 51. In *Hemings þáttr Áslákssonar* the chain of causation is set out still more succinctly. 'Síðan bjóst til Íslandsferðar. . . . Ok sem þeir Oddr létu í haf, rekr á fyrir þeim storma ok myrkr ok hafvillur, ok þekktu sik eigi, fyrr en þeir kómu at Norégi nærri Niðarósi' (ibid. vii, 435).

53. *Í.F.*, iv, 246.

54. *Ísl. sög.* xi, 186, 411.

55. She was bound from Ireland to Denmark. 'En veðr bægði honum allt vestr fyrir Írland. Síðan fær hann mikil veðr, ok stækkar sjóinn, ok þá missa þeir alla sýn af löndum ok rekr vestr í haf. Ok at lyktum þá þeir hafvillur ok vita aldri, hvar þeir eru' (ibid. iii, 453).

56. *Í.F.* iv, 246.

57. *Ísl. sög.* iv, 51.

58. *Fms.*, iii, 181. *Fas.*, iii, 402.

59. *Fas.* iii, 330. In this last case the *þokur ok hafvillur* was followed by heavy gales and disaster. Other instances of *hafvilla* followed by heavy weather may be cited from *Eiríks saga rauða* and *Finnboga saga*. 'Síðan létu þeir í haf, ok er þeir váru í haf, tók af byri. Fengu þeir hafvillur, ok fórst þeim ógreitt um summarit. . . . Sjó tók at stæra, ok þolðu menn it mesta vás ok vesold á marga vega.' 'Kippa þeir nú upp akkerum ok sigla í haf, en þá er þeir höfðu siglt nökkur dægr, tekr af byri, ok gerir á fyrir þeim hafvillur, ok vita þeir eigi, hvar þeir fara. Þar kemr, er haustar ok stærir sjóinn.' See *Fb.* i, 332; *Ísl. sög.* ix, 294–5.

60. *Ísl. sög.* xi, 186, 411.

61. Ibid. iii, 432. There are further such cases of *þokur* somewhat later in the same saga: see ibid. 450, 468.

62. *Fms.* vii, 32; *Bibliothek der angelsächsischen Poesie*, ed. C. W. Grein (Berlin, 1857), ii,

352; *Orosius*, ed. H. Sweet (1883), pp. 17–18; Adam of Bremen, *Gesta Hammaburgensis Ecclesiae Pontificum*, iv, 39–40. See also Reuter, *Germanische Himmelskunde*, p. 729; Schnall, op. cit., p. 182.

63. *Fms.* x, 112; *Maríu saga*, ed. C. R. Unger (Christiania, 1862), p. 7; Grein, op. cit. ii, 351.

64. *Ksk.*, p. 28; *Monumenta historica Norvegiae*, ed. Storm, p. 93; *Bps.* iii, 487. Cf. *Lb.*, p. 186; *Sturls.* ii, 259; *Gísli saga*, c. 49.

65. It is more than questionable whether the theory advanced by Dr. Winter to the effect that the *nauticus gnomon*, referred to by Olaus Magnus in his *Historia de gentibus septentrionalibus*, p. 343, was the instrument used by mariners in the North for 'taking' the sun, is really justified by the evidence. A careful study of the whole passage in the 1555 edition of the *Historia*, paying particular attention to the marginalia, leads one rather to the conclusion that this *nauticus gnomon* was, perhaps, some kind of azimuth compass. That the *nauticus gnomon* and the compass are one and the same instrument would appear to be pretty clearly indicated by the marginal gloss printed opposite this passage, viz. *Compassus nauticus* (no reference here to any gnomon). See also p. 66 of the *Historia*. It is not surprising to find that in the English translation of this work (1658) there is no mention at all of the gnomon, but only of the mariner's compass. Cf. Walther Vogel in *Hansische Geschichtsblätter* (1911), p. 25; Joseph Kulischer in *Allgemeine Wirtschaftsgeschichte*, vol. i, p. 306; Heinrich Winter in *Hansische Geschichtsblätter* (1937), p. 173, and in *Marine Rundschau* (1937), p. 207.

For the controversy which centred around the discovery, near the Siglufjörd in southern Greenland, of a damaged oaken disk, alleged to be some kind of bearing dial, or polaris, see Sölver, *Vestervejen om Vikingernes Sejlads*, pp. 84, 110–14. It is perhaps worth noting that on the same site there was found 'a length of braided withies or roots' remarkably similar to those discovered by the German Quay in Bergen. But see also Schnall, op. cit., pp. 85–8.

For the *sólarsteinn* controversy, see Thorkild Ramskou (Oslo, 1969), p. 59 *et passim*; Schnall, op. cit., pp. 92–115.

For the theory that a vessel's *solbyrðin*, or 'sun-boards', were some kind of device for measuring the height of the midday sun, by giving a rough indication of the vessel's northing or southing, see Falk, op. cit., 19, 54; Reuter, op. cit., p. 598 sq. But see also Schnall, op. cit., p. 90.

66. See *Lb.*, pp. 34–5; *D.I.* iii, p. 51 n.; *Rímbegla* (Copenhagen, 1780), p. 482. See also Falk, *Altnordisches Seewesen*, p. 16; Schnall, op. cit., pp. 130–4.

67. *Det gamle Grønlands beskrivelse av Ívar Bárðarson*, ed. Jónsson, p. 2; Gathorne-Hardy, *The Norse Discoverers of America*, p. 196.

68. Raold Morcken, 'Norse Nautical Units and Distance Measurements', *M.M.*, vol. 54, p. 397. See also Schnall, op. cit., pp. 130–4.

69. *Ísl. sög.*, xi, 186, 411.

70. See S. E. Morison, *Admiral of the Ocean Sea*, i, 244; D. Burwash, *English Merchant Shipping, 1460–1540* (Toronto, 1947), p. 9. Whatever may be the merits of the theory favoured by certain academic authorities that the ocean voyages of Christopher Columbus were achieved, in effect, by skilful dead reckoning, it clearly cannot apply to the ocean voyages of the Norsemen. Morison, for example, has asserted that the only instrument indispensable for Columbus was the mariner's compass; and that the other indispensable aid to navigation was a collection of sea charts. But the Norsemen, over a period of several centuries, possessed neither one nor the other. Moreover, the navigational conditions obtaining in the southern part of the North Atlantic were very different from those in the northern. Morison himself has to admit that Columbus was singularly fortunate in regard to surface currents: in that nearly all his ocean passages were made in regions of the Atlantic where the current was a negligible factor.

The fact is, that anything like regular ocean navigation, based on dead reckoning, and on dead reckoning alone, is in practice wholly impossible. The overwhelming consensus of opinion among professional seamen (as the present writer ascertained after exhaustive inquiry) is unfavourable in the highest degree to such a theory. Which must accordingly be rejected—however eminent the historian who advances it.

71. For the practical limitations of D.R., see Sølver, 'Leiðarsteinn', *Old Lore Miscellany*, vol. 10, part vii, pp. 309 sq., 320, and *Imago Mundi* (Copenhagen, 1951), pp. 212, 216 sq.; S. T. Lecky, *Wrinkles in Practical Navigation*, 1914 ed., pp. 183–5.

72. See Schnall, op. cit., p. 182. In this connection it is worth noticing that the general consensus of opinion among ornithologists appears to suggest the hypothesis that the navigation of *birds* is largely dependent upon the meridian altitude of the sun: see L. H. Matthews, 'Recent Developments in the Study of Bird Navigation', *Journal of the Institute of Navigation*, vol. 6, p. 269.

73. Morison, op. cit., i, 410. Cf. Lieut. J. W. McElroy, U.S.N.R., in *American Neptune*, vol. I, p. 212: 'The eastward and westward sailing tracks selected by Columbus were in my opinion not based upon scientific observation or secret knowledge, but were a combination of good fortune, better judgment, and the best traditions of navigation. Just like dead reckoning shipmasters down to the present century, the Admiral sought the parallel upon which his destination was presumably located and steered along that line until he reached his objective.'

74. *Í.F.* ii, 9–11; *Fb.* i, 539; *Alfræði íslensk*, i, 23; ii, 48, 72, 91; *Ísl. sög.* ix, 20–1, 377–405; *Symbolae ad geographiam medii aevi*, ed. Werlauff, p. 31; *Rímbegla*, p. 472; *Grágás*, ed. Finssen, *Konungsbók* (1852), p. 27; *Staðarhólsbók* (1879), p. 33. Cf. *Ísl. sög.* iii, 280; *Fms.* iv, 381; xi, 438; *Fld.* iii, 398; *Ksk.*, p. 5 *et passim*. Cf. also *The Book of Marco Polo*, ed. and trans. Sir Henry Yule (1903), ii, 382.

75. See *Ksk.*, p. 5.

76. *Fas.* iii, 159.

77. The Norsemen, in particular, had made such ventures long before their earliest ventures to Iceland.

78. *Orosius*, pp. 17–18. Cf. Reuter, op. cit., pp. 14–15, 725, Pl. 85; Schnall, op. cit., pp. 42, 181–2.

79. *Í.F.* iv, 246. Cf. *Ksk.*, p. 5.

80. *Fas.* iii, 195.

81. Grein, op. cit. ii, 351; *Maríu saga*, pp. 1, 7; *Fms.* x, 112. It has been suggested that the Norsemen also used the 'South Star', Wega, for the purpose of latitude determination. See *Alfræði íslensk*, ii, 72; Reuter, op. cit., p. 721. See also Schnall, op. cit., pp. 181–3.

82. *Lb.*, pp. 34–5; Reuter, op. cit., p. 728. It will be seen from the chart that the course from Bergen 'due west for Hvarf' cannot possibly pass within twelve 'sea-miles' (approximately one degree of latitude) of the south Iceland coast. The explanation may perhaps be due to an error in transcription. Cf. *Lb.*, pp. 34–5 and *Hb.*, p. 4 for the discrepancy in the distance between Reykjanes and Jolduhlaup as stated by Sturla and Hauk respectively.

83. *Hb.*, p. 4.

84. It is to be observed that if one were to steer due west (magnetic) from Hernar today, one would certainly never arrive at Hvarf. The variation of the compass in the *Hauksbók* era is unknown: but it is most improbable that throughout the later Middle Ages, when the mariner's compass was in vogue in the North, variation was a negligible factor.

85. *Ísl. sög.* vi, 46–51. Similarly, in the *Egils saga* Egil Skallagrímsson and his crew sailed from Vík with a fair wind and made their northing before they squared away to the westward, with a good slant for the south of Iceland (ibid. ii, 251). See also i, 421; ii, 218–19, 432, 456, 465; iii, 102–4; v, 342, 433; vii, 298; viii, 103–12; ix, 84; x, 223–4, 346; xi, 82, 413–14; *Í.F.* xxvii, 215; *Bps.* i, 380, 407, 464; ii, 295; *Fms.* ii, 190, 197, 280, 343. Cf. Reuter, op. cit., p. 728; Vigeland, op. cit., pp. 36, 39–41, 82.

86. Larson, *The King's Mirror*, pp. 86, 93 sq., 99 sq., 162. See Reuter, op. cit., pp. 684, 696.

87. F. Nansen, *In Northern Mists* (1911), i, 308. See Reuter, op. cit., pp. 595 sq.

88. James Fisher, author of *The Fulmar* (1952), once informed the present writer that during the Second World War, Mr. (now Sir) E. M. Nicholson was travelling to America in

one of the 'Queens' and, when the vessel was in the vicinity of the Denmark Strait, astonished her master by telling him where they were. The explanation was that Nicholson—an experienced ornithologist—had arrived at a tolerably accurate estimate of their northing by observing the appearance of a large number of the dark forms of the fulmar. The bird change in the Denmark Strait, on account of the convergence of the Gulf Stream and the colder waters into which it penetrates is particularly noticeable; and this is still more pronounced off the Newfoundland Banks. The observation of pelagic or oceanic birds in the vicinity of water convergences has to be considered in conjunction with other factors: namely, the difference in the colour of the water, the appearance in calm weather of a ripple, and, on occasion, the presence of fog.

89. *Fær. s.*, p. 104; *Hb.*, p. 443; *Bps.* ii, 290. Cf. *Ísl. sög.* iii, 432: 'Klettar vóru í fjöllum, en jöklar at ofan. þar var grægð sela ok sjódýra ok mikill fjöldi haffugls'. See also Winther, op. cit., p. 176; Fisher, op. cit., pp. 300, 312–13, 324–5.

90. *Hb.*, p. 4. The 'birds and whales' served as a useful check on their course when sailing for Hvarf. If they got too far to the southward there was of course danger of missing Greenland altogether. See Sölver, 'Leiðarsteinn', p. 319; Vigeland, op. cit., p. 39.

91. Private information.

92. For references to *hafnarmerki* in early poems, see *De gamle Eddadigte*, ed. Jónsson, p. 202; *Bps.* ii, 421; *Fas.* i, 412; Bugge, *Den norske sjøfarts historie*, p. 122. The use of landmarks, or *mið*, as a guide to the fishing-grounds is of immense antiquity in the North: see *Lb.*, p. 186. 'In the waters around the Vestmannaeyjar are to be found not less than 140 different *mið*, each with its own particular name' (*Stedsnavn*, ed. Magnus Olsen, Oslo, 1939, p. 63). Cf. *Bps.* iii, 487; *Fas.* ii, 110.

93. *Í.F.* xxvii, 253.

94. It is to be noted that on the return passage from Iceland ships occasionally made Hernar, though Stad was the more usual landfall.

95. 'The setting in or cessation of a groundswell, or a change in the colour of the water, will sometimes notify as to whether the ship is on or off soundings. Round our own coasts there are numerous well-known and strongly marked tide races, such as those off Portland and South Stack (Holyhead), which have often proved a good guide in thick weather' (Lecky, op. cit., p. 306). 'The edge of soundings may generally be recognized in fine weather by a turbulent sea and the sudden alteration in the colour of the water from dark blue to green' (Hydrographic Dept., Admiralty, *West Coast of England Pilot* (1948), p. 1). Cf. *Admiralty Weather Manual* (1941), p. 305; J. S. Learmont, *Mast in Sail* (1950), p. 101.

96. *Ísl. sög.* xi, 186, 411–12. Cf. Hydrographic Dept., Admiralty, *North Sea Pilot*, pt. I (1949), p. 237, and Chart No. 2180.

97. *Ksk.*, p. 28.

98. 'Fóru þeir svo nær Íslandi, at þeir sáu jöklana' (*Ísl. sög.* i, 428); 'sáu þeir þá til Grænlandsjökla' (ibid. 429).

99. Familiarity with these adventitious aids still formed part of the mental equipment of many an old sea-captain in the last days of sail, as may be seen from the journals and reminiscences of those times. When Captain Rex Clements first went to sea as an apprentice in the barque *Arethusa* he served under an old master who was a typical specimen of a type of navigator that has long vanished from the seas. 'Frequently in the dog watch the old man called one or other of us up on the poop and talked to us about our profession and duties. On such occasions he always used to emphasize the importance of learning to know the face of the sky and studying the ways of birds and fishes. He himself was no great reader, but his knowledge of winds, weather and natural history was truly wonderful. He navigated the ship more, I verily believe, by the look of the sea, the flight of birds, the appearance of clouds and a certain indefinable sixth sense emanating from his vast knowledge of all these, than by the more orthodox method of scientific instruments' (Rex Clements, *A Gypsy of the Horn*, 1924, p. 49).

100. The Icelandic sagas and certain other sources reveal some knowledge of surface currents: see *Fær. s.*, *p. 100; Ksk.*, *p. 35*.

101. The *Konungs Skuggsjá* has a good deal to say about the species of whales and seals in the Iceland and Greenland seas: see *Ksk.* pp. 15–17, 28–9. For the movements of whales in relation to the drift of the ice off the West Greenland coast, see R. Kellogg in the *Smithsonian Report* (1926).

102. *Í.F.* iv, p. 246.

103. See *Ísl. sög.* iv, 52. Cf. *Bps.* ii, 292–3; *Sturls.* ii, 496; *Fb.* ii, 457. For other references to pilots and pilotage, see Jónsbók, p. 241; Larson, op. cit., pp. 93, 157; *Bps.* ii, 201, 292–3; iii, 245; *Fms.* iv, 301; vii, 52; ix, 233, 376; x, 120; ix, 123; *Ísl. sög.* v, 237; vi, 17; viii, 104; *Fb.* i, 64, 145, 539; ii, 258, 457; *Fas.* iii, 113, 402; *Alexanders saga*, ed. Unger, p. 160; *Fornsögur Suðrlanda*, ed. G. Cederschiold (Leipzig, 1860), p. 105.

104. L. F. Salzman, *English Trade in the Middle Ages* (1931), p. 242.

105. J. H. Parry, *Europe and a Wider World* (1949), p. 19.

106. J. E. T. Harper, 'Navigation', *Encyclopaedia Britannica* (1951).

107. The discovery of Svalbard is noted in the Skálaholt, and also in five other Icelandic annals (*Isl. Ann.*, p. 181 *et passim*). It would appear from the *Landnámabók* that there was traffic—probably walrus-hunting—to the newly discovered country from Iceland in the thirteenth century (*Lb.*, pp. 34–5).

108. The part played by *personal knowledge and experience* in the navigation of the Norsemen can scarcely be set too high. See *Ísl. sög.* iv, 51; *Bps.* i, 484; *Sturls.* ii, 496–7; *Fb.* ii, 457. Cf. Chaucer's description of the Shipman in his *Canterbury Tales*: see *The Complete Works of Geoffrey Chaucer*, ed. Skeat (1894), iv, 12.

109. Note that the oral tradition of courses is from time to time mentioned or implied in the *Grænlendinga saga* of the Norse voyages to North America. 'We must not forget that these old societies were founded upon the spoken word to an extent which we can hardly realise': see Brøgger, *Ancient Emigrants*, p. 143. Oral instructions in all probability formed the foundation of certain interesting fragments which have survived: see *D.I.* iii, no. 9; *Det gamle Grønlands beskrivelse*, pp. 18 *sq.*

110. A relic of old Norse days may be the belief formerly held by the fishermen of the Shetland Islands that there was a significant motion of the waters—known as the 'mother-die'—which enabled them to find their way to land with nothing else to guide them. See *Saga-Book*, vol. III, p. 8.

111. *Hb.*, p. 5. See also Torfæus, *Historiae Rerum Norvegicarum* (Copenhagen, 1711), iv, 345.

112. G. J. Marcus, 'Hafvilla: A note on Norse navigation', *Speculum*, vol. 30, pp. 604–5.

113. *Lb.*, pp. 34–5; *Hb.*, p. 4. See also Reuter, op. cit., p. 728.

114. Some idea of what might be achieved by very simple methods of nautical astronomy without the aid of quadrant, astrolabe, or similar instrument can be gathered from the sailing directions set out in *Mohît*. See Maximilian Bittner, *Die topographischen Capitel des Indischen Seespiegel Mohît* (Vienna, 1897), *passim*; Joaquim Bensaude, *L'Astronomie nautique* (Berne, 1912), p. 77; Reuter, op. cit., p. 617; Sölver, *Imago Mundi*, p. 263. The navigation of the ancient Polynesians apparently reveals a knowledge of the same principle: see Thor Heyerdahl, *The Kon-Tiki Expedition* (1952), p. 152.

115. *Hakluyt Society*, Ser. i, vol. 57, p. 59; vol. 86, p. 171; vol. 99, p. 190; *C.P.S.* (*Venice*), i, 890; Richard Hakluyt, *Principal Navigations* (1903 edn.), vii, 2; McElroy, op. cit., p. 209; Sölver, *Imago Mundi*, pp. 129, 216 *sq.*, 246, 265; *M.M.*, vol. 4, p. 177; vol. 24, pp. 403–4, 408.

16. The eclipse of Norwegian shipping

1. Helgi's sons, Adalbrand and Thorvald, discovered the 'New Land' (*Isl. Ann.*, p. 142); 'Land was discovered to the west of Iceland' (ibid., p. 337); 'King Eirík sent Rólf to Iceland

to search for the New Land. . . . Rólf went out to Iceland and called for volunteers for the New Land venture (p. 384). The 'New Land', which was formerly conjectured to have been America, is now believed to have been some part of the East Greenland coast.

2. *Scriptores Rerum Danicarum medii aevi*, ed. Langebek, v, 352.

2. *Chronica Majora*, ed. H. R. Luard (1880), v, 36.

4. Knut Gjerset, *The History of the Norwegian People* (1932), i, 422.

5. It has been argued by Professor Schreiner and others that the basic cause of Norway's decline in the fourteenth century was not so much the Black Death as the changed economic situation in the north which prevented the land from recovering from the heavy loss of population. Henceforward Norwegian farming had to fit into the Hanseatic system of commerce and was thereby prevented from recovering from the effects of the devastation. See Johan Schreiner, *Hanseatene og Norge i det 16. Århundre* (Oslo, 1941), pp. 30–1, and also 'Pest og prisfall i senmiddelalderen', *Det norske videnskaps-akademie* (1948), Hist.-fil. klasse, pp. 90, 122–3, and also *Historisk Tidsskrift*, Bd.36, pp. 259–64. Cf. Maria Wetki, 'Studien zum Hanse-Norwegen-Problem', *Hansische Geschichtsblätter* (1951), pp. 43, 62–3, 66; J. A. Gade, *The Hanseatic Control of Norwegian Commerce in the Late Middle Ages* (1951), p. 9 *et passim*.

6. It has been estimated that about 40,000 craft and 300,000 persons were engaged in the herring fishery of Skania during the months of September and October. These activities centred round the towns of Skanør and Falsterbo, where the markets were held. See Bugge, *Handel og Byliv nord for Alperne*, p. 170; Gade, op. cit., pp. 16–19.

7. The cog, which appears to have been evolved by the Frisians, embodied closed cabins at quite an early date. See W. Vogel, 'Zur nord-und westeuropäischen Seeschiffahrt im fruheren Mittelalter', *Hansische Geschichtsblätter* (1907), pp. 190–1, and also *Geschichte der deutschen Seeschiffahrt* (Berlin, 1915), pp. 93–4; John J. McCusker, 'The Wine Prise and Mediaeval Mercantile Shipping', *Speculum*, Vol. 41, pp. 286–7. See also Ellmer, *Frumittel alterliche Handelsschiffahrt*, pp. 190, 266.

8. Brøgger and Shetelig, *The Viking Ships*, pp. 236–7; Schreiner in *Historisk Tidsskrift*, Bd.36, pp. 259–61; Bugge, *Den norske sjøfarts historie*, pp. 349–50, 357; Gade, op. cit., p. 67.

9. E. R. Daenell, *Die Blütezeit der deutschen Hanse* (Berlin, 1906), ii, 309; *Den norske sjøfarts historie*, pp. 249 sq., 287, 291, 357; *Hanseatene og Norge i det 16. Århundre*, pp. 20 sq., 30; Bernt Lorentzen, *Gård og grunn i Bergen i Middelalderen* (Bergen, 1952), pp. 153–6; Gade, op. cit., pp. 48–50, 53–4; P. Dollinger, *The German Hansa* (London, 1970), p. 258.

10. Thorsteinsson, *Enska öldin í sögu íslendinga*, pp. 90–3; Gade, op. cit., p. 55; *Hanseatene og Norge i det 16. Århundre*, p. 29.

11. *Isl. Ann.*, p. 207.

17. The first English voyages to Iceland

1. Richard Hakluyt, *Principal Navigations* (1903 edn.), i, 303; F. Blomefield, *An Essay towards a Topographical History of the County of Norfolk* (1775), viii, 104; Carus-Wilson, 'The Iceland Trade', p. 156; Thorsteinsson, 'The Iceland Trade', *Saga-Book*, vol. 15, p. 67, and in *Enska Öldin í sögu Íslendinga*, p. 25. The best and fullest account in English of the traffic to Iceland is Professor Carus Wilson's 'The Iceland Trade' cited above. But the substance of this essay was written as far back as 1925; and on the purely maritime side her paper is decidedly weak.

2. *Isl. Ann.*, p. 285.

3. Ibid., p. 290.

4. See Thorsteinsson, 'Íslands- og Grænlandssiglingar Englendinga á 15 öld. og 'fundur Norður-Ameríku', pp. 15–16, and also *Enska Öldin í sögu íslendinga*, pp. 27–31. See also Jón Jóhannesson, *Íslendinga saga*, ii, 156.

5. *Rotuli Parliamentorum*, iv, 79.

6. *The Libelle of Englysche Polycye*, ed. G. Warner (Oxford, 1926), p. 41; Bugge, *Handelen*

mellem England og Norge indtil begyndelsen af det 15de aarhundrede (Christiania 1898), pp. 125–9, and also *Den norske sjøfarts historie*, pp. 267–70; Gade, *The Hanseatic Control of Norwegian Commerce during the Late Middle Ages*, pp. 94–5, 98.

7. Gade, pp. 36, 37; Schreiner, *Hanseatene og Norge i det 16 Århundre*, p. 29.

8. *Icelandic Sagas*, ed. and trans. Dasent, iv, 430–1.

9. Ibid. 431.

10. *Isl. Ann.*, p. 208.

11. Ibid.; *Statutes of the Realm*, 31 Edw.III.3.c.i; *Foedera*, iii, 977; *C.C.R.*, Henry V, i, 7; *C.P.R.*, 1416–22, p. 89. Cf. *C.C.R.*, 1381–5, p. 261; 1396–9, p. 446; Henry V, i, 316, 392; *C.P.R.*, 1436–41, pp. 41–2; 1446–52, p. 479; *Black Book of the Admiralty*, ed. T. Twiss (1876), i, 273.

12. *M.M.*, vol. 5, pp. 16–17; vol. 10, pp. 95, 215; Burwash, *English Merchant Shipping*, pp. 82 *sq.*

13. *The Complete Works of Geoffrey Chaucer*, ed. Skeat, iii, 214; *H.R.* I. iv, no. 201; *H.U.* viii, no. 1160, pp. 708–9; x, pp. 292 n., 425; *Kammereirechnungen der Stadt Hamburg*, ed. K. Koppmann, ii, 14, 130; *Commento di Francesco da Buti* (1862), iii, 363; *Das Seebuch*, ed. K. Koppmann (Bremen, 1876), *passim*; Burwash, op. cit., p. 6. See also G. J. Marcus, 'The Mariner's Compass: its influence upon navigation in the later Middle Ages', *History*, N.S., vol. 41, pp. 16–24.

14. It is interesting to notice that the 'Master Mariner' in the fifteenth century *Knyghthode and Bataile* closely resembles the 'Shipman' in the *Canterbury Tales*. See *Knyghthode and Bataile*, ed. Dyboski (1935), p. 102:

> The Maister Marynere, the gouernour,
>   He knoweth euery coste in his viage
> And port saluz; and forthi grete honour
>   He hath, as worthi is, and therto wage.
> The depper see, the gladder he; for rage
>   Of wynde or of bataile if ther abounde,
>   The surer he, the ferre he be fro grounde.
>
> He knoweth euery rok and euery race,
>   The swolewys [whirlpools] & the starrys, sonde [sound] & sholde [shoal].

15. The earliest recorded case of a quadrant carried in an English vessel was in the year 1533. It would appear that the new methods of navigation were stoutly resisted by mariners of the old school. William Bourne, the author of a celebrated treatise on navigation entitled *The Regiment of the Sea* records that they 'that wer auncient masters of shippes hath derided and mocked them' who practised the new methods—'And when they dyd take the Latitude they would call them starre shooters, and would aske if they had striken it' (E. G. R. Taylor, *Tudor Geography, 1485–1583*, 1930, p. 160).

16. Hydrographic Dept., Admiralty, *Arctic Pilot*, ii (1949), pp. 62 *sq.*

17. Dasent, op. cit., p. 435. Many years ago the present writer was told by the skipper of a Hull trawler that in May 1951 there was a heavy north-easterly gale, with snow, which also began in the early morning, like this Easter gale recorded in the Icelandic annals. The most dangerous gales from this quarter are those which occur on the west coast, because wind and current are there opposed.

18. *Isl. Ann.*, p. 292; Dasent, op. cit., p. 433.

19. *D.N.* xx, no. 733. Norway had been united to Denmark, in 1397, by the Treaty of Kalmar to the detriment of Norwegian interests generally and those of the overseas dependencies in particular.

20. Ibid., nos. 736–7, 742, 755; *C.C.R.*, Henry V, i, 297; *Den norske sjøfarts historie*, p. 264.

21. *Isl. Ann.*, p. 290; *D.I.* iv, nos. 331, 377; *D.N.* xx, no. 748. See also Thorsteinsson, 'Henry VIII and Iceland', p. 68; *Hanseatene og Norge i det 16 Århundre*, p. 32. For information about the relative values of various English and Icelandic products, see *D.I.* xvi, no. 8 *et*

passim. The well-known 'stockfish tariff', set out in *D.I.* iv, no. 337, has been ascribed to this date, about 1420, but is believed by modern Icelandic historians to belong to a rather later period. See Thorkell Jóhannesson, 'Skreiðarverð á Íslandi fram til 1550', *Afmælisrit til Thorsteins Thorsteinssonar* (1950), pp. 188–94.

22. *D.I.* iv, no. 330; Gjerset, *The History of Iceland*, p. 247.

18. The English trade with Iceland

1. Warner, op. cit., p. 41.
2. Carus Wilson, 'The Iceland Trade', pp. 163, 168 *sq.*; *D.I.* iv, *passim*; xvi, pt. I *passim*; Thorsteinsson, *Enskar heimildir um sögu íslendinga a 15. og 16. öld* (1969), pp. 50–5.
3. Enskar peinildis, pp. 55–61; *D.I.* xvi, pt. I *passim*.
4. Congregation Books, i, fos. 84, 200–1, 262–3; Carus Wilson, 'The Iceland Trade', p. 174; Thorsteinsson, 'Siglingar til Íslands frá Biscups Lynn', *A góðu dægri: afmæliskveðju til . . . Sigurðar Nordals* (1951), pp. 5, 8–9, and also *Enska öldin í sögu íslendinga*, pp. 101–5.
5. *D.I.* xvi, nos. 26, 28, 31; Carus Wilson, 'The Iceland Trade', p. 173. Cf. *Letters and Papers of Richard III and Henry VII*, p. 287.
6. P.R.O.: E.122/19/1; *C.P.R.*, 1446–52, pp. 137, 156, 191, 528; C.1/43/277–8; *D.N.* xx, no. 856; Carus Wilson, 'The Iceland Trade', p. 169.
7. *C.P.R.*, 1436–41, pp. 70–1; *D.I.* xi, no. 15; xx, no. 859; *Statute Rolls . . . of Ireland, Henry VI*, p. 697; Carus Wilson, 'The Iceland Trade', pp. 168–9.
8. 'Before the foresail was regularly carried it is highly probable that to keep the ship steady before the wind her boat's sail was sometimes set on the fore flagstaff, or on the boat's mast stepped temporarily on the forecastle and that a two-master was thus gradually evolved before men realized that they were changing their rig': see Alan Moore, *The Development of Sail-Plan in Northern Europe* (London, 1926), p. 5.
9. P.R.O.: E.101/53/11; *Naval Accounts and Inventories of the Reign of Henry VII*, ed. Michael Oppenheim (1896), *passim*; Burwash, *English Merchant Shipping*, pp. 82 *sq.*; *M.M.*, vol. 5, pp. 16–17; vol. 10, pp. 95, 215.
10. No adequate definition of the various ship types is possible on the basis of the evidence at present available. The word ship, or *navis*, was often applied to larger vessels in a general sense. The carrack was a Mediterranean vessel, the special characteristic of which was its immense size. The hulk was of Baltic origin. The barge was a sailing craft of about 70 tuns burden, which may or may not have carried oars as an auxiliary means of propulsion. The balinger was a good deal smaller, averaging some 50 tuns, and was fitted with auxiliary oars. The crayer, which was rather smaller, apparently carried no oars. Barges and balingers were especially common in the West Country. Crayers, like doggers, were often used as fishing-boats. The English ship types commonly mentioned in connection with ocean voyages were ships, balingers, and doggers.
11. See Burwash, op. cit., p. 87.
12. *Archaeologia*, vol. 57, i (1899), pp. 30 *sq.*; Burwash, op. cit., pp. 82 *sq.*; Laird Clowes, *Sailing Ships: Their History and Development*, pp. 52 *sq.*
13. Moore, op. cit., pp. 4–8.
14. *Pageant of the Birth, Life, and Death of Richard Beauchamp, Earl of Warwick*, ed. Dillon and St. John Hope (1914), *passim*; R. and R. C. Anderson, *The Sailing Ship* (1926), pp. 119–22.
15. The whole question of tunnage, or tonnage, in the medieval period is a difficult and highly controversial one. One of the best surveys of the evidence is that set out by Dorothy Burwash in her *English Merchant Shipping*, pp. 88–95. See also McCusker, 'The Wine Prise and Mediaeval Mercantile Shipping, p. 286.
16. Carus Wilson, 'The Iceland Trade', p. 240.
17. Burwash, Op. cit., pp. 96, 100.

18. E. M. Carus Wilson, *The Overseas Trade of Bristol in the Later Middle Ages* (Bristol, 1937), pp. 71–3, 79–80, 125, 130, 135.

19. *Registres des grands jours de Bordeaux, 1456, 1459*, ed. H. A. Barckhausen (Paris, 1867), pp. 422–6; *Enskar heimildir*, pp. 32–57.

20. Carus Wilson, 'The Iceland Trade', p. 167, and also *The Overseas Trade of Bristol*, pp. 81, 144, 147–8, 252–3; Thorsteinsson, 'Islands-verzlun Englendinga', *Skírnir* (1951), p. 103.

21. See Marcus, 'The Mariner's Compass', pp. 16–24.

22. Compasses were also carried, as we know from Olaus Magnus's *Historia*, in Norwegian fishing-vessels working far out in the Atlantic: see Olaus Magnus, *Historia de gentibus septentrionalibus* (Rome, 1555), p. 730.

23. Salzman, *English Trade in the Middle Ages*, pp. 242–4; Carus Wilson, 'The Iceland Trade', p. 160. Alwyn Ruddock repeats this error in 'John Day of Bristol and the English Voyages across the Atlantic before 1497', *Geographical Journal* (1966), p. 20.

24. *The Cely Papers*, ed. H. E. Malden (1900), p. 177.

25. P.R.O.: H.C.A. High Court of Admiralty 24, File 5, large bundle, John Aborough *contra* John Andrews.

26. *Cock Lorrell's Boat*, ed. Roxburghe Club (1817). It was well understood that the lodestone might perhaps be used to falsify the compass needle. See Olaus Magnus, op. cit., p. 343: 'Qui . . . malitiose nauticam gnomonen, aut compassum, & praecipue portionem magnetis, vnde omnium directio dependet, falsauerit', etc. That lodestones were carried on board ship until well on in the eighteenth century can be seen from the interesting collection preserved in the National Maritime Museum, Greenwich.

27. *Admiral of the Ocean Sea*, i, 245.

28. British Library: Lansdowne MS. 285, fos. 139–40; *Itineraria Symonis Simeonis et Willelmi de Worcestre*, ed. J. Nasmith (1778), *passim*.

29. Nasmith, *passim*.

30. Lansdowne MS. 285, loc. cit.

31. *D.I.* xvi, no. 149; *Enska öldin*, p. 153.

32. Congregation Books, i, fos. 262–3; *D.I.* xvi, nos. 88, 89; *D.N.* xx, nos. 777, 778, 800, 818; Carus Wilson, 'The Iceland Trade', pp. 165–7; *Enskar heimildir*, pp. 92–4.

33. Congregation Books, i, fos. 115, 120, 247–8; *Statutes of the Realm*, 8 Henry VI, cap. 2; 10 Henry VI, cap. 3; *Rotuli Parliamentorum*, iv, 378; *D.N.* xx, no. 776; Carus Wilson, 'The Iceland Trade', pp. 166 *sq.*

34. *C.P.R.*, 1422–9, p. 394; Bugge, *Den norske sjøfarts historie*, p. 303.

35. P.R.O.: Treaty Roll, 126 m.13 *et passim*; *D.N.* xx, nos. 812, 816, 817, 819, 825, 847; Carus Wilson, *The Overseas Trade of Bristol*, pp. 71–3, 78–81.

36. Warner, op. cit., p. 41. Cf. *Statute Rolls . . . of Ireland, Henry VI* (1910), p. 697. In connection with this matter of empty holds it is worth noticing that the tradition once existed in Hull that Hedon Market Place, and perhaps other parts of the town as well, were paved with stone slabs imported from Iceland. 'At such tyme as al the trade of Stokfish for England cam from Iseland to Kingeston, bycause the burden of stokfisch was light, the shipes were balissid with great coble stone brought out of Iseland, the which yn continuance pavid al the town of Kingeston throughout.' That this would not be the first time that such weighty goods were brought from overseas to Hull is evident from the customs accounts for the year 1400–1, which record the arrival in Hull of two Dutch vessels, the *Mariknight* of Amsterdam, with 40,000 'pavyng-stones' included in her lading, and the *Shankewyn* of Dordrecht, with 16,000. See Leland, *Itinerary*, ed. L. Toulmin Smith (1907), i, 61; Chas. Frost, *Notices relative to the early history . . . of Hull* (1827), App. 15, 17; Thorsteinsson, 'Islandsverzlun Englendinga', p. 106.

37. *D.N.* xx, no. 794. Cf. no. 800.

38. These were the *Cuthbert* of Berwick, the *Clement* of Hedon, and the *Holy Ghost* of Schiedam: see E.122/61/32. There was also the *Christopher* of Hull (master, John Hall) which

returned with a cargo of stockfish, train-oil, and wadmal: see P.R.O. Memoranda Rolls, 159/207, Mich. 1430–Trin. 1431.

39. C.P.R., 1436–41, pp. 70–2; 1441–6, pp. 191, 479; Carus Wilson, 'The Iceland Trade', pp. 168 sq., and also The Overseas Trade of Bristol, pp. 65–8, 87–91.

40. P.R.O.: E.122/18/39 passim; E.122/61/32 passim; C.P.R., 1436–41, p. 294; Carus Wilson, 'The Iceland Trade', pp. 175–6, and also The Overseas Trade of Bristol, p. 208.

41. British Library Cotton MS, Vespasian E.ix. Cf. D.I. iv, no. 381; Carus Wilson, 'The Iceland Trade', pp. 175–6; Jón Jóhannesson, Íslendinga saga, ii, 192–6.

42. D.I. iv, no. 381; D.N. i, nos. 670, 756; v, no. 585; xx, nos. 776, 800, 820, 832, 837; C.P.R., 1436–41, pp. 234–5; P.R.O. Treaty Rolls, 125–6 passim.

43. Caspar Weinreichs Danziger Chronik, ed. Hirsch and Vossberg (1885), p. 4; H.U. ix, 326–7; Gjerset, History of Iceland, pp. 265–6; Thorsteinsson, 'Henry VIII and Iceland', Saga-Book, vol. 15, p. 71, and also Enska öldin, pp. 204–22.

44. D.I. vi, no. 426; vii, no. 112.

45. D.I. vi, nos. 66, 67; x, no. 27; xii, no. 26; xvi, nos. 25, 27; H.R. II. vii, no. 348; Carus Wilson, 'The Iceland Trade', pp. 180–2; Enskar heimildir, pp. 53, 60.

46. H.R. III. i, no. 511, p. 558. See also D.I. xi, no. 43.

47. D.I., xvi, no. 26 et passim; E. Baasch, Die Islandfahrt der Deutschen (Hamburg, 1899), p. 58; Thorsteinsson, 'Íslandsverzlun Englendinga', p. 95, and also 'Henry VIII and Iceland', p. 75.

19. The Iceland fishery

1. Bodleian Douce MS. 98, fos. 195–6.

2. See Thorsteinsson, Enskar heimildir, p. 45; and also Enska öldin, p. 51.

3. D.I. iv, no. 381.

4. Hydrographic Dept., Admiralty, Arctic Pilot, ii, 106 sq., 120, and Chart No. 565.

5. 'Þat veer hofum saet up hws i vestmannaøyum uttan nokors løfre': see D.N. xx, no. 749.

6. D.I. iv, no. 381. See also F. Magnusen, Om de Engelskes Handel og Færd paa Island i det 15de Aarhundrede (Copenhagen, 1833), p. 119; Carus Wilson, 'The Iceland Trade', pp. 164–5; Enska öldin, pp. 61–4.

7. D.I., iv, no. 381. The names of several of the leading offenders (for example, John Percy, John Pasdale, Robert Thirkell, Thomas York, and Rawlin Beck) are mentioned in the Bench Books preserved in the Guildhall, Kingston upon Hull. See Bench Books, ii, 246 sq.; Enska öldin, pp. 75–6.

8. See Gjerset, History of Iceland, p. 261.

9. See Thorsteinsson, 'Sendiferðir og hirðstjórn Hannesar Pálssonar og skýrsla hans 1425', Skírnir (1953).

10. Isl. Ann., p. 274; Enska öldin, p. 71.

11. D.I. iv, no. 386.

12. Letters and Papers of Richard III and Henry VII, ii, 287.

13. The Paston Letters, ed. J. Gairdner (Edinburgh, 1910), vi, 136–7.

14. Bugge, Den norske sjøfarts historie, p. 306; Carus Wilson, 'The Iceland Trade', p. 172; Thorsteinsson, 'Henry VIII and Iceland', pp. 73–4.

15. T. Gardner, An historical account of Dunwich, anciently a city (1754), pp. 145, 147, 248.

16. Thorsteinsson, 'Henry VIII and Iceland', p. 74.

17. L.P.F.D. iii, nos. 3071, 3248; iv, no. 5101; Addenda, no. 873; P.R.O.: High Court of Admiralty 24, File 5, no. 12. See G. J. Marcus, 'The English Dogger', M.M., vol. 40, pp. 294–6; Thorsteinsson, Enska öldin, p. 260, and also 'Íslands- og Grænlandssiglingar Englendinga á 15. öld og fundur Norður-Ameríku', Saga (1969), p. 13.

18. Baasch, Die Islandfahrt der Deutschen, p.58.

19. Andrew Borde, *The Fyrst Boke of the Introduction of Knowledge*, ed. Viles and Furnivall (1879), pp. 141–2.

20. P.R.O.: E.122/99/16–23.

21. V. B. Redstone, *The Ancient House* (Ipswich, 1912), pp. 99–100; J. A. Webb, *Great Tooley of Ipswich* (Ipswich, 1962), p. 72 *et passim*.

22. *Ibid.*, p. 103.

23. E. R. Cooper, 'The Dunwich Iceland Ships', *M.M.*, vol. 25, p. 173.

24. *L.P.F.D.* iii, nos. 3071, 3248. See also Thorsteinsson, 'Henry VIII and Iceland', pp. 72–3.

20. The English rivalry with the Hanse

1. *H.R.* I. vi, no. 262; *D.I.* iv, no. 384; *D.N.* xx, no. 789. See also P. Dollinger, *The German Hansa* (London, 1970), pp. 242–3.

2. P.R.O.: E.122/61/32; *C.P.R.*, 1436–41, p. 270; *D.I.* viii, no. 41. The Dutch, however, were heavily handicapped in the Iceland traffic by the fact that they needed to import grain themselves.

3. *H.R.* II. i, no. 393, pp. 318–19. Cf. no. 294.

4. Thorsteinsson, 'Henry VIII and Iceland', pp. 72–3.

5. Ibid., p. 98.

6. *H.U.* ix, no. 519; Gade, *The Hanseatic Control of Norwegian Commerce*, pp. 90–2; Dollinger, op. cit., pp. 306–7.

7. *Kammereirechnungen der Stadt Hamburg, 1471–1500*, ed. Koppmann, iii (1878), pp. 223, 253; Baasch, *Die Islandfahrt der Deutschen* (Hamburg, 1899), pp. 6–8; Thorsteinsson, *Enska öldin*, pp. 224–5.

8. *H.R.* iii, i, no. 501; nos. 54, 351, 510; *H.U.* ix, nos. 686, 800; *D.I.* vi, nos. 362, 605; viii, no. 357; x, no. 30; xi, nos. 22, 30, 32, 41, 49; Bruns, op. cit., pp. lxv–lxvii, 213; Hirsch and Vossberg, *Caspar Weinreichs Danziger Chronik*, p. 23; Baasch, *Die Islandfahrt der Deutschen*, pp. 8–10, 95, 104; Bugge, *Den norske sjøfarts historie*, p. 307; Schreiner, *Hanseatene og Norge i det 16. Århundre*, p. 43; Gade, op. cit., p. 113; Thorsteinsson, *Enska öldin*, pp. 226–8. The Hanse imported *váðmal*, wool, butter, and tallow from the Faeroe and Shetland Islands: see *H.R.* iii, vi, no. 613, art. 8.

9. *D.N.* vi, no. 609; *D.I.* vii, no. 12; viii, no. 72; x, nos. 22, 31; xi, nos. 23, 30, 34, 52, 54; *H.U.* x, no. 1201; *L.P.F.D.*, Addenda, no. 873; Bruns, op. cit., pp. 212–3; *Hamburgische Chroniken in niedersächsischer Sprache*, ed. J. M. Lappenberg (Hamburg, 1861), p. 302; Baasch, op. cit., pp. 8, 58, 103; Thorsteinsson, 'Henry VIII and Iceland', pp. 81, 98–9, and *Enska öldin*, pp. 212–22.

10. For the crucial importance of the cog in the seaborne commerce of the Hanse, see Pagel, *Die Hanse* (Oldenburg, 1943), pp. 254–5; Schreiner in *Norsk Historisk Tidsskrift*, Bd. 36, pp. 259–60; Gade, op. cit., p. 64.

11. Vogel, *Geschichte der deutschen Seeschiffahrt*, pp. 467–70, 496; *Den norske sjøfarts historie*, pp. 285–9; Pagel, op. cit., pp. 257–8.

12. *H.U.* ix, nos. 95, 122–3, 127; *H.R.* II. vi, nos. 538, 551; Max Lehrs, *Der Meister W A* (1895), ff. xi–xiii; B. Hágedorn, *Die Entwicklung der wichtigsten Schiffstypen bis ins 19. Jahrhundert* (Berlin, 1914), p. 63; Vogel, op. cit., pp. 466, 475, 481, 496–7; R. C. Andersen, *The Sailing Ship*, p. 119; Pagel, op. cit., p. 254.

13. Baasch, op. cit., pp. 103–4; Bugge, *Den norske sjøfarts histoire.*, p. 293; R. Morton Nance, 'A Hanseatic Bergentrader of 1489', *M.M.*, vol. 3, p. 161; Vogel, op. cit., pp. 466, 496–7; Sölver, 'Danske Skibstegninger fra det 15. Aarhundrede', *Vikingen*, no. 12. Cf. Laird Clowes, *Sailing Ships: Their History and Development*, p. 56. Dr. Heinrich Winter has some interesting observations on this subject in his pamphlet, *Hanse-Kogge um 1470* (Berlin, 1946).

14. Gade, op. cit., pp. 64–5.

15. Lappenberg, op. cit., p. 169. Some idea of the duration of a normal Iceland voyage

can be gathered from the fact that in 1475 a Hamburgman was lying off the west coast of Iceland as early as the month of April. See *D.I.* x, no. 27; Baasch, op. cit., pp. 95–6.

16. *H.R.* I. iv, no. 201; *H.U.* viii, no. 1160, pp. 708–9; x, p. 292 n.; *ibid.*, p. 425; *Kammereirechnungen der Stadt Hamburg*, ii, 130.

17. Kratschmer, *Die italienischen Portolane des Mittelalters* (Berlin, 1909), pp. 198–9.

18. *Das Seebuch*, ed. Koppmann, *passim*; *H.U.* viii, no. 21, p. 10. See also Vogel, op. cit., pp. 522–7.

19. *Safegarde of Saylers, or great Rutter*, trans. Robert Norman (1590), ff. 13a–14b.

20. Lappenberg, op. cit., p. 109; *D.I.* x, no. 27; Baasch, op. cit., pp. 95–6.

21. *H.R.* II. i, p. 111.

22. *C.S.P.* (*Milan*), 1385–1618, no. 548. See also Hirsch and Vossberg, op. cit., pp. 12, 25.

21. The last ventures to Greenland

1. *G.H.M.* iii, 125, 139–40; *D.N.* xviii, no. 33; Jóhannesson, *Íslendinga saga*, ii, 338–54; Gad, *Grønlands historie*, i, 156–8.

2. Gad, pp. 156–8; Thorsteinsson, *Enska öldin*, pp. 33–4. Morison writes: 'A ship from Norway spent a winter at the Western Settlement between 1406 and 1410; no people were there, only cattle running wild' (*European Discovery*, p. 36). This is a serious error; the ship from Norway actually called at the *East* Settlement, and spent not one winter there, but several. The West Settlement, as is clearly established, had become extinct in the early half of the previous century: to which era Ívar Bárdarson's well-known reference to the deserted homesteads and the cattle running wild belongs.

3. It is not without significance that the large stone halls in Greenland were a product of the later medieval era: see Shetelig, 'Den norrøne byggeskikk på Grønland', *Naturen*, no. 6, p. 167.

4. *Isl. Ann.*, p. 288; *G.H.M.* iii, 148–9.

5. *G.H.M.* iii, 148–9.

6. *Isl. Ann.*, p. 289.

7. *D.N.* xvii, no. 759; Nørlund, *Viking Settlers in Greenland*, p. 144; Jóhannesson, op. cit., ii, 353–4.

8. Thorsteinsson is of opinion that the men of Bristol had probably found their way to Greenland early in the fifteenth century; and had sailed there from time to time whenever the wind was fair for the passage. See *Enska öldin*, p. 291.

9. Thorsteinsson, 'Íslands- og Grænlandssiglingar Englendinga og fundur Norður-Ameríku á 15. öld', pp. 28–33, 49, and also *Enska öldin*, pp. 32–5, 268–9.

10. *G.H.M.* i, 126–7; 'Íslands- og Grænlandssiglingar Englendinga og fundur Norður-Ameríku á 15. öld', p. 26, and also *Enska öldin*, pp. 105–6.

11. *D.N.* xx, no. 794.

12. Nørlund, 'Buried Norsemen at Herjólfsnes', *M.o.G.*, Bd.67, pp. 149 *sq.*, 181 *sq.*, 221, 236, 251 *sq.*, and *Viking Settlers in Greenland*, p. 125; Gad, op. cit., pp. 188–9.

13. Gad, p. 342; Brøgger, *Vinlandsferdene* (Oslo, 1937), pp. 168 *sq.*

14. See *Enska öldin*, pp. 291–2.

15. Nørlund, 'Buried Norsemen at Herjolfsnes', pp. 66, 219, and *Viking Settlers in Greenland*, p. 342. The theory propounded by Dr. Hansen to the effect that malnutrition was responsible for the extinction of the colony, that the Greenlanders became physically degenerate, and that few of them attained middle age is now largely discredited. This theory is seen to have been based on insufficient evidence; and careful examination of the skeletal remains on other sites has produce very different conclusions to those arrived at nearly fifty years ago by Hansen. See F. C. C. Hansen, 'Anthropologia medico-historica Groenlandiae antiquae', *M.o.G.*, Bd.67, pp. 518 *sq.*

16. *Det gamle Grønlands beskrivelse*, ed. Jónsson, pp. 29–30.

17. See Therkel Matthiassen, 'Archaeology in Greenland', *Antiquity*, vol. 9, p. 195.

18. Thorsteinsson, 'Some Observations on the Discoveries of the Norsemen', *Saga-Book*, vol. 16, p. 180.

19. *G.H.M.* iii, 121–2.

20. *Isl. Ann.*, p. 364.

21. H. Rink, *Eskimoiske Eventyr og Sagn* (Copenhagen, 1866), pp. 198–209.

22. Gad, op. cit. i, 195–7.

23. See Thorsteinsson, *Enskar heimildir*, pp. 92–4, and *Enska öldin*, p. 106.

24. *D.N.* xx, no. 800.

25. *D.N.* vi, no. 527. See Ingstad, *Land under the Pole Star*, pp. 329–30.

26. *Land under the Pole Star*, pp. 326–9. Cf. W. H. Graah, *Undersøgelser-Reise til Ostkysten af Grønland* (Copenhagen, 1832), p. 7.

27. Ostermann, 'Niels Egedes Beskrivelse over Grønland, *M.o.G.* Bind. 120, pp. 48–9. For another remarkable example of Eskimo tradition, in this case relating to the Elizabethan explorer, Martin Frobisher, see Morison, *European Discovery*, p. 526.

28. *G.H.M.* iii, 513 *sq.*

29. Christian Vebæk, *Beretninger vedrørende Grønland*, no. 1, sect. ix, p. 96, and also 'Vatnaherfi', *Fra Nationalmuseets Arbejdsmark* (1952), p. 114.

22. The English voyages of discovery

1. It has been suggested by G. R. Crone that the discovery in the later medieval era of the Madeiras, Azores, Canaries, and Cape Verde Islands had revived interest in legendary islands such as St. Brendan's Isle and the Isle of Brasil : see G. R. Crone, 'The Mythical Islands of the Atlantic Ocean', *Comptes Rendus du Congrès International de Geographie* (Amsterdam, 1938), vol. 2, sect. IV, p. 167.

2. *Itineraria Symonis Simeonis et Willhelmi de Worcestre*, ed. Nasmith, p. 153.

3. P.R.O., E. 19/16; *Memoranda Roll*, 22 Edw. IV, Hilary m. 30; *Memoranda Roll*, 22 Edw. IV, Hilary m. 10.

4. *C.S.P. (Sp.)*, I, 1485–1509, no. 210. See *The Cabot Voyages and Bristol Discovery under Henry VII*, ed. J. A. Williamson (1962), pp. 26, 62, 114, 202; Quinn, *England and the Discovery of America, 1481–1620* (London, 1973) pp. 14, 85–6; Patrick McGrath, 'Bristol and America, 1480–1631', *The Westward Enterprise*, ed. K. R. Andrews, N. P. Canny, and P. G. H. Hair (Liverpool, 1978), pp. 83, 86.

5. Archivo General de Simancas: Estado de Castilla, leg. 2, fol. 6; L. A. Vigneras in *Canadian Historical Review*, vol. 38 (1957), p. 227.

6. *The Cabot Voyages and Bristol Discovery under Henry VII*, ed. Williamson, pp. 6, 30–1 ; Thorsteinsson, 'Íslands- og Grænlandssiglingar Englendinga og fundur Nordur-Ameriku a 15. old', pp. 64, 69; Quinn, op. cit., p. 28.

7. Íslands- og Grænlandssiglingar Englendinga', pp. 20–1 ; Thorsteinsson, *Enska öldin*, pp. 262–5.

8. Cf. Alwyn Ruddock in *Geographical Journal* (1966), p. 331. It is always to be borne in mind that the various theories which have grown up around this problem (including, of course, the present writer's) are based in large measure upon surmise and conjecture. Unless and until further evidence comes to light, therefore, there will be no knowing which of these explanations comes nearest to the truth.

9. See C. V. Sölver and G. J. Marcus in *M.M.*, vol. 44, pp. 20–1, 33–4.

10. Thorsteinsson, *Enska öldin*, p. 43 *et passim*.

11. Williamson, op. cit., pp. 114–15; Quinn, op. cit., p. 105.

12. Vigneras, op. cit., p. 228.

13. Ibid., p. 227; *C.S.P. (Milan)*, no. 552. See Williamson, op. cit., p. 62.

14. Vigneras, op. cit., p. 227; *C.S.P. (Milan)*, i, no. 552; Williamson, op. cit., p. 208. The identification of Cabot's landfall and subsequent cruise along the neighbouring coast presents so many problems that it has given rise to endless speculation and surmise. Despite

the late J. A. Williamson's wise warning that the whole matter should be kept strictly objective, it inevitably tends to become subjective. Morison writes: 'The reader should keep in mind that English and Anglo-Canadian historians are desperately eager to prove that Cabot touched the American mainland, so that they can claim a "first" for him as discoverer of the continent, Columbus not having set foot on the mainland before 1498' (*European Discovery*, p. 193). This rather naïve suggestion reflects quite as much on Morison himself as on the historians he has in mind. In any case, it is not patriotic sentiment which casts doublt on the claims formerly made for Columbus as 'first across': but the evidence, especially the latest evidence.

15. Vigneras, op. cit., p. 227.

16. Ibid., pp. 227–8.

17. H. P. Biggar, *The Precursosrs of Jacques Cartier, 1497–1534* (Ottawa, 1911), p. 29.

18. *C.S.P.* (*Milan*), i, no. 552.

19. Biggar, op. cit., pp. 14–15.

20. Williamson, op. cit., pp. 221, 225.

21. Ibid., pp. 72–3.

22. Ibid., pp. 118–38; Quinn, op. cit., p. 130. See also Alwyn Ruddock, 'The Reputation of Sebastian Cabot', *Institute of Historical Research Bulletin*, vol. 47 (1974), pp. 96–8.

23. W. G. Gosling, *Labrador: its discovery, exploration, and development* (London, 1910), *passim*. There is an excellent account of the French ventures to the Newfoundland fishery in *European Discovery*, pp. 263–6, 272–3.

24. Biggar, op. cit., p. 136; Gosling, op. cit., pp. 35–6; Thorsteinsson, 'Íslands- og Grænlandssiglingar og fundur Norður-Ameríku á 15. öld', p. 70.

25. Richard Hakluyt, *Principal Navigations* (1589), p. 720. Cf. Sir George Peckham on the Voyage of Sir Humphrey Gilbert: 'For after once we are departed the coast of England, we may passe straight way thither, without danger of being driuen into any the countryes of our enemies, or doubtfull friends: for commonly one winde serueth to bring vs thither, which seldome faileth from the middle of Januarie to the middle of May, a benefite which the mariners make great account of, for it a pleasure that they haue in few or none of other iourneyes' (ibid., p. 714). See also A. B. Becher, *Navigation of the Atlantic* (London, 1892), p. 145.

Glossary

THE DEFINITIONS ARE for the most part derived from William Falconer's *Universal Dictionary of the Marine* (1769) and Admiral William Smyth's *A Sailor's Word-Book* (1878).

Abaft. This word, generally speaking, means behind, inferred relatively, beginning from the stem and continuing towards the stern, that is, the hinder part of the ship.

Adrift, To be (Old Norse: *at reka*). Floating at random, at the mercy of the wind, seas, or current, or all of them together.

Altitude. The elevation of any of the heavenly bodies above the plane of the horizon, or its angular distance from the horizon, measured in the direction of a great circle passing through the zenith.

Amidships. The middle of the ship, whether in regard to her length between stem and stern, or in breadth between the two sides.

Anchor (Old Norse: *akkeri*). A large and heavy instrument in use from the earliest times for holding and retaining ships, which it executes with admirable force. With few exceptions it consists of a long iron shank, having at one end a ring, to which the cable is attached, and the other branching out into two arms, with flukes or palms at their bill or extremity. A stock of timber is fixed at right angles to the arms, and serves to guide the flukes perpendicularly to the surface of the ground.

Anchorage (Old Norse: *skipalægi*). Ground which is suitable, and neither too deep, shallow, or exposed for ships to ride in safety upon.

Arming. A piece of tallow placed in the cavity and over the bottom of a sounding lead, to which any objects at the bottom of the sea become attached, and are brought with the lead to the surface.

Ballast (Old Norse: *grjót*). Heavy material such as stone deposited in a vessel's hold when she either has no cargo or too little to bring her sufficiently low in the water.

Bank (Old Norse: *rif*). A rising ground in the sea, differing from the shoal, because not rocky but composed of sand, mud, or gravel.

Beat to windward (Old Norse: *beita undir veðrit*). The operation of sailing obliquely to windward, in a zigzag line, or traverse.

Blinder (Old Norse: *blindsker*). Sunken rock.

Bow. The fore-end of a ship or boat.

Bowline. A rope used to keep the weather edge of the sail tight forward and steady when the ship is close-hauled to the wind.

Bowsprit. A large spar projecting over the stem.

Broach-to. Fly up into the wind.

Burgoo (Old Norse: *grautr*). Boiled oatmeal mixed with butter.

Búza (Old Norse). A trading-ship which in the later Middle Ages largely supplanted the *knörr* on the shorter Atlantic routes, but never on the Greenland venture.

Cable (Old Norse: *akkeris-strengr*). A thick, strong rope which serves to keep a ship at anchor.

Carry away. Break.

Carvel-built. A vessel or boat, the planks of which are all flush and smooth, the edges laid close to each other, and caulked to make them water-tight: in contradistinction to clincher-built, where they overlap each other.

Cog (Old Norse: *kuggr*). A bluff-bowed, broad-beamed, roomy cargo vessel in general use on the Hanse trade-routes.

Curach (Old Irish). A hide-covered craft of wickerwork, propelled by sail and oars, used by the Irish in the early medieval period.

Current (Old Norse: *straum*). A certain progressive flowing of the sea in one direction, by which all bodies floating therein are compelled more or less to submit to the stream. The *setting* of the current is that point of the compass towards which the waters run; and the *drift* of the current is the rate it runs at in an hour.

Dead reckoning. The estimate of a ship's position from the courses steered and distance run.

Deep-waterman (Old Norse: *hafsiglingarmaðr*). A mariner engaged in ocean voyages.

Departure. The bearing of an object on the coast from which a vessel commences her dead reckoning and takes her departure.

Fair wind (Old Norse: *byrr*). That which is favourable to a ship's course, in opposition to contrary or foul.

Fog (*Old Norse: þoka*). A dense mist at sea.

Founder (Old Norse: *kafa*). Sink.

Freeboard. The space between the gunnel and the line of flotation.

Gale (Old Norse: *storm-viðri*). A wind of considerable strength, but not so strong as a storm.

Garboard strakes. The first range of planks laid upon a ship's bottom, next to the keel.

Goosewing (Old Norse: *svidris vid raa midia*). The clews or lower corners of a sail when the middle part is furled or tied up to the yard.

Ground-tackle (Old Norse: *grunn-færi*). A general name given to all sorts of ropes and fittings appertaining to the anchors, or employed in mooring or otherwise securing a ship in a roadstead or harbour.

Gunnel (Old Norse: *skipsbord*). A piece of timber going round the upper sheer-strake as a binder for its top-work.

Halyard (Old Norse: *dragreip*). The rope employed to hoist or lower the sail upon its yard.

Haven (Old Norse: *havn*). A safe refuge from the violence of wind and sea; much the same as harbour, though of less importance. A good anchorage rather than place of perfect shelter.

Haze. A greyish vapour, less dense than a fog.

Head-wind (Old Norse: *andviðri*). A wind which will not permit a ship to sail her course; a dead muzzler.

Helm (Old Norse: *stjörn*). Properly is the tiller, but sometimes used to signify the rudder, and the means used for turning it.

Hour-glass. The sand-glass: a measure of the hour.

Hull (Old Norse: *húfr*). The body of a ship, independent of masts, yards, sails, and rigging.

Ice-blink. A streak or stratum of lucid whiteness which appears over the ice in that part of the atmosphere adjoining the horizon.

Ice-floes (Old Norse: *ísjaki*). Fields of floating ice of any extent, as beyond the range of vision.

Iron-bound (Old Norse: *öræfi*). A coast where the shores are composed of rocks which mostly rise perpendicularly from the sea, and have no anchorage near to them.

Jacob's staff, or *Cross-staff*. A mathematical instrument to take altitudes, consisting of a brass circle, divided into four equal parts by two lines cutting each other in the centre; at each extremity of either line is fixed a sight perpendicularly over the lines, with holes below each slit for the better discovery of distant objects. The cross is mounted on a staff or stand for use.

Keel. The lowest and principal timber of a ship, running fore and aft its whole length, and supporting the frame like the backbone in quadrupeds.

Keelson. An internal keel, laid upon the middle of the floor timbers, immediately over the keel, and serving to bind all together.

Knörr (Old Norse). A broad-built, sturdy sailing -vessel which was used by the Norsemen in the navigation to Iceland and Greenland.

Landfall (Old Norse: *sýn af landi*). The first land discovered after a passage.

Lead. An instrument for discovering the depth of water, attached to the lead line, which is marked at certain distances to ascertain the fathoms.

Leeward. The direction to which the wind blows.

Leeway. What a vessel loses by drifting to leeward in her course.

Lubber. An awkward unseamanlike fellow.

Lull. The brief interval of moderate weather between the gusts of wind in a gale. Also, an abatement in the violence of surf.

Magnetic needle. A balanced needle, highly magnetized, which points to the magnetic pole, when not influenced by the local attraction of neighbouring iron.

Mark (Old Norse: *mið*). Bearings of objects on shore to mark the fishing banks.

Master (Old Norse: *stýrimaðr*). The mariner in command of a vessel.

Meridian. An imaginary great circle passing through the zenith and the poles, and cutting the equator at right angles. When the sun is on the meridian of any place, it is midday there, and at all places situated under the same meridian.

Mirage (Old Norse: *hillingar*). Literally, the 'upheaving', optically, of objects such as islands and mountains from their true location, enabling them to be observed at an abnormal distance.

Moorings (Old Norse: *landfestar*). The means of securing a vessel in a particular station by chains or cables.

Navigator. A person skilled in the art of navigation.

Northing. In navigation, implies the distance made good towards the north: the opposite of southing.

Overfall (Old Norse: *áföll*). A rippling or race in the sea, where, by the peculiarities of bottom, the water is propelled with immense force, especially when the wind and tide, or current, set strongly together.

Parallel sailing. Sailing nearly on a given parallel of latitude.

Pilot (Old Norse: *leiðsagnarmaðr*). The mariner in charge of the navigation of a ship.

Pole-star (Old Norse: *leiðarstjarna*). This most useful star is the lucida of the Little Bear, round which the other components of the constellation and the rest of the heavens appear to revolve in the course of the astronomical day.

Port (Old Norse: *bakborði*). The distinguishing term for the left side of a ship when looking forward; the opposite of starboard.

Pump (Old Norse: *dæluastr*). A machine for discharging water from the bottom of the ship's hold.

Quadrant. A reflecting instrument used to take the altitude above the horizon of the sun, moon, or stars at sea, and thereby to determine the latitude of the place.

Race (Old Norse: *röst*). Strong currents producing overfalls, dangerous to small craft. They may be produced by narrow channels, crossing of tides, or uneven bottoms.

Reef. Reduce sail in proportion to the increase of wind.

Rigging (Old Norse: *skip-reiði*). A general name given to all ropes employed to support the masts and work the sails.

Seamanlike (Old Norse: *skipkænn*). Showing the skill of a good seaman.

Seamanship. The art of managing a ship at sea.

Sea-mark (Old Norse: *hafnarmark*). A point or object distinguishable at sea, such as promontories, beacons, etc.

Sea-way (Old Norse: *sævargangr*). A rough sea.

Shipwreck (Old Norse: *skipsbrot*). The destruction of a vessel by her driving on rocks, sands, etc., either on the shore or off-shore, as distinct from her foundering in deep water.

Shoal (Old Norse: *grunn*). A term synonymous with shallow, signifying that part of the water not sufficiently deep to sail over by vessels which are navigated in that place; a shallow, however, is never supposed to be dry, even at the lowest ebb, but shoals are often dry at low water.

Shrouds (Old Norse: *höfuðbenda*). The lower and upper standing rigging.

Slant. A wind of which advantage may be taken.

Sounding. The operation of ascertaining the depth of the sea, and the quality of the ground, by means of a lead and line, sunk from a ship to the bottom, where some of the ooze or sand adheres to the tallow in the hollow base of the lead.

Soundings. In common parlance signifies parts of the ocean not far from the shore where the depth is about 80 or 100 fathoms.

Spring-tide. The periodical excess of the elevation and depression of the tide, which occurs when both the sun and moon act in the same direction.

Squall (Old Norse: *byrl*). Sudden gust of wind.

Stay (Old Norse: *stag*). A large strong rope, extending from the upper end of the mast towards the stem of the ship, as the shrouds are extended on each side. The object of both is to prevent the masts from springing, when the ship is pitching deep. Thus stays are fore and aft; those which are led down to the vessel's stern are *back-stays*.

Steep-to (Old Norse: *sjóbrattr*). A steep coast where the water, deepening rapidly, admits the near approach of shipping without the danger of grounding.

Stem (Old Norse: *fremstafn*). The foremost timber forming the bow of a vessel.

Stern (Old Norse: *skutr*). The after-part of a ship or boat.

Strake. One breadth of a plank in a ship, either within or without board, wrought from the stem to the stern-post.

Trenails. Long cylindrical wooden pins employed to fasten the planks of a ship's side and bottom to the corresponding timbers.

Variation. A term applied to the deviation of the magnetic needle or compass, from the true north point toward either east or west.

Weather (Old Norse: *stýra a veðr*). To weather anything is to go to windward of it.

Weather-wise (Old Norse: *veðrkænn*). Skilful in forecasting the weather. Judging from the appearance of the sea and sky, the behaviour of birds, etc., the gyrations of cobwebs, even, some mariners would attain to an almost magical prescience of a coming shift of wind or bad weather.

Weigh (Old Norse: *draga upp grunn-færi*). Set a ship in motion by raising the anchor.

Windward. The direction from which the wind blows.

Winter nights (Old Norse: *vetrnætr*). The three days and nights which began the winter season on the northern sea-routes.

Zenith. The pole of the horizon, or that point in the heavens directly overhead.

Bibliography

Manuscript Sources

Archivo General de Simancas: Estado de Castilla, leg. 2, fol. 6.
Bodleian Library, Oxford; Douce MS. 98
British Library, London: Lansdowne MS. 285
Kingston upon Hull: Bench Books, II and III
Lynn: Congregation Books, I and II
Pepysian Library, Cambridge: Sea Manuscripts, 2934
Public Record Office, London: Early Chancery Proceedings
——: Treaty Rolls
——: Exchequer, K.R., Customs Accounts
——: Exchequer, K.R., Memoranda Rolls
——: High Court of Admiralty
Yale University: 'Vinland Map'

Printed Sources

Adam of Bremen, *Gesta Hammaburgensis Ecclesiae Pontificum*, ed. Lappenberg, J. M. (Hanover, 1876)
Ailred of Rievaulx, *The Life of St. Ninian*, ed. Forbes (Edinburgh, 1874)
Anderson, W. B. (ed.), *Sidonius: Poems and Letters* (London, 1936)
Annals of Ulster ed. W. Hennessy (Dublin, 1887)
Atkinson, R. (ed.), *The Book of Ballymote* (Dublin, 1887)
Barckhausen, H. A. (ed.), *Registres des grands jours de Bordeaux, 1456, 1459* (Paris, 1867)
Berthelot, A. (ed.), *Festus Avienus: Ora Maritima* (Paris, 1936)
Bittner, M. (ed.), *Die topographischen Capitol des Indischen Seespiegel Môhit* (Vienna, 1897)
The Book of Marco Polo, ed. and trans. Sir Henry Yule (London, 1903)
Botfield, B. (ed.), *Manners and Household Expenses of England in the Thirteenth and Fifteenth Centuries* (London, 1841)
Borde, Andrew, *The Fyrst Boke of the Introduction of Knowledge*, ed. F. J. Furnivall (EETS, 1870)
Calendar of Close Rolls
Calendar of Patent Rolls
Calendar of State Papers
Campbell, J. F. *Popular Tales of the West Highlands* (Edinburgh, 1890–3)
Carmichael, A. (ed.), *Carmina Gadelica* (Edinburgh, 1928)
Carus Wilson, E. M. *The Overseas Trade of Bristol in the Later Middle Ages* (Bristol, 1937)
Caspar Weinreichs Danziger Chronik, ed. Hirsch and Vossberg (Berlin, 1885)
Cely Papers, ed. H. E. Malden (Camden Society, Third Series, I, 1900)
Claudian, *De consulate Stilichonis*

BIBLIOGRAPHY

Collier, J. P. (ed.), *Household Books of John, Duke of Norfolk, and Thomas, Earl of Surrey, 1481–1490* (1841)

Dicuil, *Liber de mensura orbis terrae*, ed. J. J. Fierney (Dublin, 1967)

Die Lübecker Bergenfahrer und ihre Chronistik, ed. F. Bruns (Berlin, 1900)

Diplomatarium Færonense, ed. Jakob Jakobsen (Copenhagen, 1907)

Diplomatarium Islandicum, ed. Jón Sigarðsson and others (Copenhagen/Reykjavik, 1857–)

Diplomatarium Norvegicum, ed. Langebek and Urger (Christiania/Oslo, 1847–)

Early Norwegian Laws, ed. and trans. L. M. Larson (New York, 1935)

Færeyinga saga, ed. C. C. Rafn (Copenhagen, 1832)

Finsen, V. (ed.), *Grágás, Islænderines Lovbog* (Copenhagen, 1851)

Flateyjarbók, ed. G. Vigjusson and C. R. Unger (Christiania, 1860)

Fornaldar sögur Norðurlanda (Reykjavik, 1943)

Fornmanna sögur (Copenhagen, 1825–)

Gairdner and Morgan (ed.), *Sailing Directions for the Circumnavigation Of England* (1889)

Gildas, *De excidio et conquestu Britanniae*, ed. M. Winterbottom (Chichester, 1978)

Giraldus Cambrensis, *Topographia Hibernica*, ed. J. F. Dimock (RS 1867)

Grein, C. W. (ed.), *Bibliothek der angelsachsischen Poesie* (Belin, 1857)

Grønlands historiske Mindesmærke, ed. C. C. Rafn and F. Magnusson (Copenhagen, 1838)

Hakluyt, R. *The Principal Navigations, Voyages and Discoveries of the English Nation* (ed. 1589, and Glasgow, 1903)

Hamburgische Chroniken in niedersächsischer Sprache, ed. J. M. Lappenberg (Hamburg, 1861)

Hamel, A. G. Von (ed.) *Immrama* (Leyden, 1941)

Hanserecesse, ed. G. von de Roff (Leipzig, 1876–92)

Hansisches Urkundenbuch, ed. C. Höhlbaum and others (Halle and Leipzig, 1876–1916)

Heist, W. W. (ed.), *Vitae Sanctorum Hiberniae ex Codice olim Salmanticensi nunc Bruxellensi* (Brussels, 1965)

Hermannsson, H. (ed.), *The Book of the Icelanders* (Ithaca, New York)

Historicae Catholica Iberniae Compendium (Lisbon, 1621)

Icelandic Sagas, ed. and trans. R. G. Dasent (London, 1894)

Islandske annaler, ed. G. Storm (Christiania, 1888)

Íslenzk Fornrit, ed. Jakob Benediktsson and others (Reykjavik, 1959)

Itineraries of William Wey (Roxburghe Club, 1857)

Jónsbók, ed. Ól. Halldórsson (Copenhagen, 1904)

Jónsson, F. (ed.), *De gambe Eddadigte* (Oslo, 1932)

—— (ed.), *Den norsk-islandske skjaldedigtning* (Christiania, 1908–15)

—— (ed.), *Det gamle Grønlands beskrivelse af Ivar Barðarson* (Copenhagen, 1930)

—— (ed.), *Hauksbók* (Copenhagen, 1892)

—— (ed.), *Landnámabók*, (Copenhagen, 1900)

Jónsson, G., *Byskupa sögur* (Reykjavik, 1950)

——, *Islendinga sögur* (Reykjavik, 1947)

——, *Sturlunga saga* (Reykjavik, 1944)

Julius Caesar, *De bello civile*

——, *De bello gallico*

Keating, *History of Ireland*, trans. D. Comyn and P. S. Dineen (Dublin, 1908)

Knyghthode and Bataile, ed. R. Dybroski and L. M. Arend (EETS, 1935)

Koffmann, K. (ed.), *Das Seebuch* (Bremen, 1876)

Konungs Skuggsja, ed. L. Holm-Olsen (Oslo, 1949)

Kammereirechnungen der Stadt Hamburg, ed. K. Koffmann (Hamburg, 1878)

Larson, L. M. (ed.), *The King's Mirror* (New York, 1917)

Laxdæla saga, trans. T. Veblen (New York, 1925)

Leland, John, *Itinerary*, ed. L. Toulmin Smith (1907)

Letters and Papers Foreign and Domestic of the Reign of Henry VIII, ed. J. Gairdner (1861–3)

Letters and Papers of Richard III and Henry VIII

BIBLIOGRAPHY

Libelle of Englyshe Polyce, ed. G. Warner (Oxford, 1926)
Life of St. Lawrence, trans. O. Elton (1890)
Mabinogion, ed. and trans. Gwyn and Thomas Jones (London, 1949)
Meddedelser om Grønland (Copenhagen, 1879–)
Meissner, R. (ed.), *Stadtrecht des König Magnus Hakonarson* (Berlin, 1950)
Monumenta historica Norvegiae, ed. G. Storm (Christiania, 1880)
Munch, P. A. (ed.), *Pavelige Nuntiers Regnskabs- og Dagbøger Jørte under Tiende Opkrævningen i Nordern, 1282–1334* (Christiania, 1864)
Nasmith, J. (ed.), *Itineraris Symonis Simeonis et Willelmi de Worcestre* (Cambridge 1778)
Neckam, *De Naturis rerum*, ed. T. Wright (London, 1883)
Nordal, S. (ed.), *Orkneyinga saga* (Copenhagen, 1913)
Norges gamle Love indtil 1387, ed. Keyser and Munch (Christiania, 1846)
Notices relative to the early history . . . of Hull, ed. Chas. Frost (1827)
O'Donovan, ed. and trans. *Annals of . . . the Four Masters* (Dublin, 1851)
O'Grady, S., ed., *Silva Gadelica (1892)*
Olaus Magnus, *Historia de gentibus septentrionalibus* (Rome, 1555)
Oppenheim, M. (ed.), *Naval Accounts and inventories of the Reign of Henry VIII 1485–1488 and 1495–1497* (1896)
Orkneyinga saga, ed. and trans. A. B. Taylor (Edinburgh, 1932)
Orosius, ed. H. Sweet (1993)
Paris, Matthew, *Chronica Majora*, ed. H. E. Luard (RS, London, 1872–83)
Paston Letters, ed. J. Gairdner (Edinburgh, 1910)
Pliny, *Historia naturalis*
Plummer, C. (ed.), *Vitae Sanctorum Hiberniae* (Oxford, 1910)
—— (ed.), *Bethada Náem nÉrenn* (Oxford, 1922)
Reeves, W. ed. and trans. *The Life of St. Columba* (Dublin, 1857)
Rímbegla (Copenhagen, 1780)
Rotuli Parliamentorum (London, 1767–77)
Rymer, T. (ed.), *Foedera* (The Hague, 1739–1745 edn.)
Safegarde of Saylers, or great Rutter, trans. Robert Norman (London, 1590)
Sagas of Kormac and the Sworn Brothers, The, ed. and trans. Lee M. Hollander (New York, 1949)
Scriptores Rerum Danicarum medii aevi, ed. J. Langebek (Copenhagen, 1772–8)
Selmer, C. (ed.), *Navigatio Sancti Brendani Abbatis* (Notre Dam, Ind., 1959)
Skeat, W. (ed.), *The Complete Works of Geoffrey Chaucer* (Oxford, 1894)
Skene, W. F. (ed.), *Chronicles of the Picts and Scots* (Edinburgh, 1867)
Solinus (ed.), *Polyhistor* (1689)
Stokes, W. (ed.), *Sanas Chormaic* (Dublin, 1868)
—— (ed.), *The Tripartitie Life of St. Patrick* (London, 1889)
The Story of Egil Skallagrimsson, trans. W. C. Green (London, 1893)
Turville-Petre, G. E. O. and Olszewska, ed. and trans. *The Life of Gudmund The Good* (Oxford, 1942)
Twiss, T. (ed.), *Monumenta Juridica: the Black Book of the Admiralty* (RS 1871–6)
Unger, C. R. (ed.), *Alexanders saga* (Christiania, 1848)
—— (ed.), *Maríu saga* (Christiania, 1862)
Vatnsdalers Saga, The, trans. Gwyn Jones (1944)
Vinssen (ed.), *Grágás* (Copenhagen, 1852)
Wallace, J. *An Account of the Islands of Orkney* (Edinburgh, 1700)
Werlauff, E. C. (ed.), *Symbolae ad geographiam medii aevi* (Copenhagen, 1821)

Dictionaries

Blondel, S. *Íslensk-Donsk Orðabók* (Copenhagen, 1924)
Cleasby and Vigfusson, *Icelandic–English Dictionary* (Oxford, 1874)
Falconer, W. *A New Universal Dictionary of the Marine* (1815 edn.)
Fitzner, *Ordbok over det gamle norske sprog* (Christiania, 1886)
Jal, A. *Archaéologie navale* (Paris, 1840)
Smyth, W. *A Sailor's Word-Book* (1878)

Periodicals

Aabøger for nordisk Oldkyndighed og Historie
Acta Archaeologica
American Neptune
Annales de Bretagne
Antiquity
Archaeologia
Beretninger vedrørende Grønland
Bergens Sjøfartsmuseum Årshefte
Canadian Historical Review
English Historical Review
Ériu
Études celtiques
Geographical Journal
Hansische Geschichtsblätter
Historisk Tidsskrift
History
Irish Ecclesiastical Record
Journal of Roman Studies
Journal of the Institute of Navigation
Kronborgmuseets Aarbog
Marine Rundschau
Mariner's Mirror
Nordisk Historia
Nordisk Kultur
Nordisk Tidsskrift
Norsk Geografisk Tidsskrift
Norsk Historisk Tidsskrift
Norsk Kulturhistorie
Proceedings of the Royal Irish Academy
Révue celtique
Saga
Saga Book of the Viking Society for Northern Research
Scottish Gaelic Studies
Skírnir
Speculum
Transactions of the Gaelic Society of Inverness
Zeitschrift für deutsches Alterthum

Other Works

Andersen, M., *Vikingefærden* (Christiania, 1895)
Anderson, R. and R. C., *The Sailing Ship: Six Thousand years of History* (London, 1927)

BIBLIOGRAPHY

Baasch, E., *Die Islandfahrt der Deutschen* (Hamburg, 1899)

Becher, A. B., *Navigation of the Atlantic* (1892)

Bensande, J., *l'Astronomie nautique au Portugal à l'epoque des grands découvertes* (Berne, 1912)

Bieler, L., *Ireland: Harbinger of the Middle Ages* (Dublin, 1963)

Biggar, *The Precursors of Jacques Cartier, 1497–1534* (Ottawa, 1911)

Blomefield, F. *An Essay towards a Topographical History of the County of Norfolk* (1775)

Borderie, A. de la, *Histoire de Bretagne* (Rennes, 1896)

Brøgger, A. W. *Ancient Emigrants* (Oxford 1929)

—— *Løgtinssøga Føroya* (Oslo, 1935)

—— *Vinlandsferdere* (Oslo, 1937)

Brøgger, A. W. and Shetelig, H. *Vikingeskipere* (Oslo, 1950)

—— *The Viking Ships* (Oslo, 1953)

Bruun, D., *Erik den Røde* (Copenhagen, 1930)

Bugge, A., *Den norske sjøfarts historie* (Oslo, 1923)

—— *Den norske trælasthandel historie* (Skien, 1933)

—— *Norse Loan Words in Irish* (Halle, 1912)

—— *Studies over de norske byers sclustyre og handel for Hanseaternes Tid* (Christiania, 1899)

—— *Vesterlandenes Indflydelse* (Christiania, 1905)

Burwash, D., *English Merchant Shipping, 1460–1540* (Toronto, 1947)

Buxton, D. (ed.), *Custom is King* (London, 1936)

Chadwick, H. M., *Early Scotland* (London, 1949)

Chadwick, N. K., *The Age of the Saints in the Early Celtic Church* (London, 1961)

——, *Early Brittany* (London, 1969)

Churchill, W. S. *History of the English-speaking Peoples* (London, 1956)

Clark, J. G. D., *Prehistoric Europe* (London, 1952)

Clements, R., *A Gypsy of the Horn* (London, 1924)

Clowes, G. S., Laird, *Sailing Ships: Their history and development* (London, 1948)

Daenell, E. R., *Die Blütezeit der deutschen Hanse* (Berlin, 1906)

Dillon, M., *Early Irish Literature* (Chicago, 1948)

Dillon, M., and Chadwick, H. K., *The Celtic Realms* (London, 1967)

Dollinger, P., *The German Hansa* (London, 1970)

Ellmer, D., *Frühmittelalterliche Handelsschiffahrt* (Neumünster, 1972)

Falk, H., *Altnordisches Seewesen* (Heidelberg, 1912)

Falk, H. and Shetelig, H., *Scandinavian Archaeology* (Oxford, 1937)

Fisher, H. A. L., *History of Europe* (London, 1957 edn)

Fisher, J., *The Fulmar* (London, 1952)

Foote, P. G., *On the Saga of the Faroe Islanders* (London, 1964)

Frere, S. S., *Britannia* (London, 1974 edn.)

Gad, F., *Grønlands historie* (Copenhagen, 1967)

Gade, J. A., *The Hanseatic Control of Norwegian Commerce in the Late Middle Ages* (Leyden, 1951)

Gaffarel, P., *Histoire de la découverte de l'Amérique* (Paris, 1892)

Gardner, T., *An Historical Account of Dunwich, anciently a city* (London, 1754)

Gathorne-Hardy, G. M., *The Norse Discoverers of America* (Oxford, 1921)

Gatty, H., *The Raft Book* (New York, 1948)

Gjerset, K., *The History of Iceland* (London, 1924)

——, *The History of the Norwegian People* (New York, 1932)

Gjessing, G., *Fangstfolk* (Oslo, 1941)

——, *Yngre Steinalter i Nord-Norge* (Oslo, 1942)

BIBLIOGRAPHY

Gosling, W. G., *Labrador: its discovery, exploration, and development* (London, 1910)
Grah, W. H., *Undersøgelser-Reise til Ostkysten af Grønland* (Copenhagen, 1882)

Hagedorn, B., *Die Entwicklung der wichtigsten Schiffstypan bis ins 19 Jahrhundert* (Berlin, 1914)
Hartshorne, C. H., *Early Reminiscences of the Great Island of Aran* (Dublin, 1853)
Hasund, S., *Det norske folks liv og historier gjennem tidere* (Oslo, 1934)
Haugen, E., *The Kensington Stone* (Wisconsin, 1932)
——, *Voyages to Vinland* (New York, 1942)
Helland-Hansen and Nansen, *The Norwegian Sea* (1909)
Henderson, I., *The Picts* (London, 1967)
Heyerdahl, T., *The Kon-Tiki Expedition* (1952)
Hitchins and May, *From Lodestone to Gyro-Compass* (1951)
Holand, Hj., *Kensington Stone* (Wisconsin, 1932)
——, *Westward from Vinland* (New York, 1940)
Hornborg, E., *Segelsjöfartens Historia* (Stockholm, 1923)
Hornell, J., *British Coracles and Irish Curraghs* (1938)
——, *Water Transport* (1946)
Hovgaard, W., *The Voyages of the Norsemen to America* (New York, 1914)
Hydrographic Dept, Admiralty, *Arctic Pilot* (1949 edn.)
——, *Irish Coast Pilot* (1930 edn.)
Hydrographic Dept, Admiralty, *North Sea Pilot* (1949 edn.)
——, *West Coast of England Pilot* (1949 edn.)
——, *West Coast of Scotland Pilot* (1949 edn.)

Ingstad, H., *Westward to Vinland* (1958)
——, *Land under the Pole Star* (1966)

Jackson, K., *Language and History in Early Britain* (Edinburgh, 1953)
Jakobsen, J., *Shetlandsøerenes stednavne* (Christiania, 1901)
——, *Dialect and Place Names of Shetland* (Lerwick, 1897)
Johannesson, W., *Gerdir Landnámabókar* (Reykjavik, 1941)
——, *Íslandingar saga* (Reykjavik, 1958)
Jones, Gwyn, *The Norse Atlantic Saga* (Oxford, 1965)
——, *The Vikings* (1969)

Kemp, P. K., *Oxford Companion to Ships and the Sea* (1976)
Kendrick, T. D., *History of the Vikings* (1930)
Kenney, J. F., *Sources for the Early History of Ireland* (Dublin, 1929)
Kermack, W. R., *Historical Geography of Scotland* (1926)
Kimble, G. H. T., *Geography in the Middle Ages* (1938)
Kingsford, C. L., *Prejudice and Promise in Fifteenth century England* (Oxford, 1925)
Kartschmer, K., *Die italienischen Portolane des Mittelalters* (Berlin, 1909)

Learmont, J. G., *Master in Sail* (1950)
Lecky, S. T., *Wrinkles in Practical Navigation* (17th edn.)
Lehrs, M., *Der Meister W. A.* (Dresden, 1895)
Lethbridge, T. C., *Merlin's Isle* (1948)
——, *Herdsman and Hermits* (1950)
Lewis, A. R., *The Northern Seas* (Princeton, 1958)
Lindquist, S., *Gotlands Bildsteine* (Stockholm, 1942)
Lorentzen, B., *Gård og grunn i Bergen i Middelalderen* (Bergen, 1952)
Loth, J., *L'émigration bretonne en Armorique* (Paris, 1883)

Magnusen, F., *Om de Engelskes Handel og Færd paa Island i det 15de Aarhundrede* (Copenhagen, 1883)

Melsted, B. T., *Ferðir, siglingar, og samgöngur milli Island og annara landa á dögum þjóðveldins* (Reykjavik, 1914)

Mohn, *Vindene i Nordsjøen ok Vikingetogen* (Christiania, 1914)

Moore, A., *The Development of Sail-Plan in Northern Europe* (1926)

Morison, S. E., *Admiral of the Ocean Sea* (Boston, 1942)

——, *The European discovery of America: the Northern Voyages* (1971)

Muller-Roder, *Die Beizjagd und der Falkensport in alter und neuer Zeit* (Berlin, 1906)

Musset, L., *Les peuples scandinaves au moyen âge* (Paris, 1951)

Nansen, F., *In Northern Mists* (1911)

Nicholas, H., *A History of the Royal Navy* (London, 1847)

Nørlund, P., *Norse Ruins at Gardar* (1930)

——, *Viking Settlers of Greenland* (1936)

Nordal, S., *Íslenzk Menning* (Reykjavik, 1942)

O'Flaherty, R., *Ogygia* (London, 1685)

O'Grady, Standish H., *Pacata Hibernia* (1896)

Olsen, M., *Farms and Fanes of Ancient Norway* (Christiania, 1928)

——, *Stedsnavn* (Oslo, 1939)

Oppenheim, M., *History of The Administration of the Royal Navy, 1509–1660* (London, 1896)

Paasche, F., *Landet med mørke skibene* (Oslo, 1938)

Pagel, H., *Die Hanse* (Oldenburg, 1943)

Parry, J. H., *Europe and a Wider World* (London, 1949)

Pokorny, *History of Ireland* (London, 1933)

Power, E. and Postan, M. M. (ed.), *Studies in English Trade in the Fifteenth Century* (London, 1933)

Ramskou, T., *Solstennen* (Oslo, 1969)

Redstone, V. B., *The Ancient House or Sparrowe House, Ipswich* (Ipswich, 1912)

Reman, E., *Norse Discovery and Explorations in America* (Berkeley and Los Angeles, 1949)

Reports of the Royal Commission on Ancient and Historical Monuments of Scotland: Orkney and Shetlands (1946)

Reuter, O. S., *Germanische Himmelskunde* (Munich, 1934)

Rink, H., *Eskimoiske Eventyr og Sagn* (Copenhagen, 1866)

Røyel, J., *The 1949 Cruise of the Viking Ship 'Hugin'* (Copenhagen, 1949)

Rygh, O., *Norske gaardnavne* (Oslo, 1924)

Salzman, L. F., *English Trade in the Middle Ages* (1931)

Sawyer, P. H., *The Age of the Vikings* (1962)

Schnall, U., *Navigation der Wikinger* (Hamburg, 1975)

Schreiner, J., *Hanseatene og Norge i det 16 Århundrede* (Oslo, 1941)

——, *Hanseatene og Norges nedgang* (Oslo, 1935)

Shetelig, H., *Préhistorie de la Norvège* (Oslo, 1926)

Shetelig, H. and Johannessen, *Kvalsundfundet og andre norske myrfund av fartier* (Bergen, 1929)

Skelton, R. A., Marston, T. E. and Painter, G. D. *The Vinland Map and the Tartar Relation* (New Haven, 1965)

Sölver, C. V., *Imago Mundi* (Copenhagen, 1951)

Sölver, C. V., *Om Ankre* (Copenhagen, 1945)

——, *Vestervejen om Vikingernes Sejlads* (Copenhagen, 1954)

Steenstrup, J., *Normannerne* (Copenhagen, 1886)

BIBLIOGRAPHY

Stefánsson, V., *Ultima Thule* (New York, 1942)
Stove, S. and Jakobsen, J., *Færoyane* (Oslo, 1944)
Sveinsson, E. Ól., *Landnám i Skaftafellsþingi* (Reykjavik, 1948)
Synge, J. M., *The Aran Islands* (1907)

Taylor, E: G. R., *The Haven-Finding Art* (1958)
——, *Tudor Geography, 1485–1583* (1930)
Thompson, J. W., *Economic and Social History of the Middle Ages* (1928)
Thórdarson, M., *The Vinland Voyages* (New York, 1930)
Thorsteinsson, B., *Enska öldin í sögu íslendinga* (Reykjavik, 1970)
——, *Íslenska þjóðveldið* (Reykjavik, 1953)
Tonning, O., *Commerce and Trade on the North Atlantic, 850 AD to 1350 AD* (Minnesota, 1935)
Torfæus, *Historiae Rerum Norwegicarum* (Copenhagen, 1711)
Turville-Petre, G. E. O., *The Heroic Age of Scandinavia* (1951)

Vigeland, N., *Norge på Havet* (Oslo, 1953)
Vogel, W., *Geschichte der deutschen Seeschiffahrt* (Berlin, 1915)

Washburn, W. E., *Proceedings of the Vinland Conference* (Chicago, 1971)
Watson, W. J., *Celtic Place-Names of Scotland* (1926)
Webb, J., *Great Tooley of Ipswich* (Ipswich, 1962)
Williamson, J. A., *Maritime Enterprise 1485–1558* (Oxford, 1913)
—— (ed.), *The Cabot Voyages and Bristol Discovery under Henry VIII* (1962)
—— (ed.), *Voyages of the Cabots* (1929)
Winter, H., *Hanse-Kogge um 1470* (Hamburg, 1946)
Winther, N., *Færøernes Oldtidshistorie* (Copenhagen, 1875)

INDEX